Nuclear Power?

Compiled by
Tom Unterrainer

Spokesman Dossier

Published in 2022 by
Spokesman
5 Churchill Park
Nottingham, NG4 2HF, England
www.spokesmanbooks.org

Spokesman is the publishing imprint of the Bertrand Russell Peace Foundation Ltd.

All rights reserved.

ISBN 9780851249032

A cataloguing-in-publication (CIP) record is available from the British Library.

Contents

Introduction

Nuclear Power: What is at stake ... 1
Malcolm Caldwell

The Politics of Nuclear Energy ... 13
Alan Roberts

The Sizewell Syndrome ... 43
Tony Benn

Neither Safe Nor Essential ... 69
Petra Kelly

Why Chernobyl Still Matters ... 80
Rosalie Bertell

Is Nuclear Power Safe? ... 97
Christopher Gifford

Fukushima ... 106
Zhores Medvedev

Fukushima's Radioactive Elements ... 109
Helen Caldicott

Fukushima's Quagmire ... 115
Hachiro Sato

Nuclear Explosions ... 119
Christopher Gifford

UK Energy Policy ... 133
Ian Fairlie

Submerged Politics of UK Nuclear Power ... 138
Phil Johnstone & Andy Stirling

Nuclear Reactors and Climate Change ... 150
Pete Roche & Ian Fairlie

Stop Trying to Make Nuclear Energy Happen ... 161
Dave Cullen

Introduction

The papers and articles collected in this *Spokesman Dossier* span five decades. As such, you might expect many of the arguments to be dated or even irrelevant in the third decade of the twenty-first century. Sadly – and a little surprisingly – this is not generally the case.

Take, for example, Tony Benn's evidence to the Sizewell Inquiry (1984). Benn argued that coal and other fossil fuels present a preferable means of securing energy supplies than would nuclear. Given what we now know, such an argument alone would be unviable, to say the least. However, this is not his central argument. In fact, Benn's evidence to the Sizewell Inquiry opens an invaluable window on the mechanisms by which sections of government and industry work together to further a complex of financial and military interests. Benn is clear on the link between nuclear power generation and the needs of British and associated nuclear weapons systems: a link still 'submerged' in general understanding of the issues, as Phil Johnstone and Andy Stirling explain in their more recent article.

The first item re-published here is by Malcolm Caldwell. It is part of a longer article on 'The Energy Crisis', published in 1972 in a remarkable collection titled *Socialism and the Environment*. Edited by Ken Coates, this volume brings together a series of papers presented to a conference organised by the German Metal-workers' Union, I.G. Metall, on 'The Quality of Life'. Caldwell dissects the claims made for nuclear energy and finds that the hopes behind the claims are "sagging, if not receding". Why this conclusion? The costs, delays, dangers and damaging environmental impact of nuclear energy production was as evident in 1972 as it is in 2022. So why do certain governments persist in this wasteful and dangerous enterprise? Why do some entertain the idea that nuclear energy has 'green credentials'?

Alan Roberts, who went on to become a campaigning Labour MP, points out in 'The Politics of Nuclear Power' (1977), that the drive towards nuclear energy generation is intimately linked with the overall dynamics of capitalism; an argument addressed again by Dave Cullen in the final essay in this *Dossier* (2021).

The stirring words of Petra Kelly in her 1986 article, 'Neither Safe Nor Essential', should have removed all uncertainty about the dangers of nuclear energy. Written shortly after the disaster at Chernobyl and delivered as an address to the Oxford Union, Kelly starkly outlines the perils presented by nuclear reactors. The world did not listen. Nearly two decades later, we have Rosalie Bertell writing that 'Chernobyl Still Matters' (2003). Still, the world ignored the warnings. By the 2010s, we have the deadly lessons of Fukushima. Still, the world ignores the warnings.

By the 2020s, not only are billions of pounds to be ploughed into new nuclear reactors, but so also is the fantasy of 'nuclear fusion' (always '25 years away') still tickling the synapses of all-too-many. We are now supposed to believe that nuclear energy generation will be the saviour of a world on the brink of climate catastrophe! The grotesque proportions of this transformation are the main motivation for producing this *Dossier* at a time when the world faces very many other dangers and acute crises.

We are supposed to forget the political-economic-military nexus driving nuclear power. We are supposed to forget Chernobyl, forget Fukushima, and forget all the other deadly nuclear incidents. We are supposed to forget about the toxic nuclear waste that will be created by a new generation of nuclear reactors. The public is supposed to believe that the billions to be spent on new nuclear reactors would not be better spent on clean, renewable, truly 'green' energy sources.

The writers collected here devoted their talents and energies to exposing the dangers posed by nuclear power. We should follow their example.

Tom Unterrainer
December 2021

Nuclear Power: What is at stake

Malcolm Caldwell

Malcolm Caldwell (1931-1978) was a prolific writer, activist and academic. He served as Chair of the Campaign for Nuclear Disarmament from 1968-1970 and dedicated his talents to many similar causes. Malcolm Caldwell was murdered during a visit to Cambodia in 1978, hours after meeting Pol Pot.

This excerpt is from 'The Energy Crisis', published in Socialism and the Environment *(edited by Ken Coates) and published by Spokesman in 1972.*

We must now turn to serious consideration of nuclear energy. The highest hopes have been reposed in this innovation since its first early stages. We believe that it is true to say that hopes are now sagging, if not receding. But the question deserves full discussion.

There are both fission and fusion reactions. The latter are the more complex, and their controlled realisation remains a pious aspiration: " ... few scientists active in the field of plasma physics today are willing to predict whether such a device can be developed to generate *useful* power within the next 25 years – and it is plasma physicists who must solve the critical problems"[59] The problems are proving more intransigent than anticipated, and we may safely relegate any prospect of successful commercial harnessing of nuclear fusion for social energy into the twenty first century - if indeed it ever proves feasible.

We may, therefore, concentrate upon nuclear fission. Fission reactors are divided into three types – burners, converters and breeders. Burner reactors, which consume the naturally occurring fissile isotope uranium-235, are subject to the objection we raise below of shortage of the raw material input. Realisation of this limitation early led to research directed towards "conversion" and "breeding". These processes, in effect, transform materials which are not themselves fissionable into previously nonexistent isotopes which are fissionable. These materials are known as "fertile".[60] "The neutrons required for conversion or breeding are those produced in a reactor whose initial supply of fuel is uranium-235. If uranium-238, or thorium-232, is placed in such a reactor, some of its atoms will absorb neutrons and become converted into its respective fissile isotope. The basic difference between conversion and breeding, is that by means of a conversion reactor, only a fraction of the fertile material can be converted into fissile material before the supply of the latter is completely exhausted. Whereas, for the breeder reactor, more fissile material is produced than is consumed, and it is possible, in principal (sic), to utilize the entire supply of fertile material, provided that sufficient uranium-235 is available to start the process initially."[61]

Practically speaking, the majority of existing or realistically projected nuclear power-plants to date have been burners. Those responsible for future power supplies are perfectly aware of what this entails in the way of raw material inputs. Milton Shaw, Director of the American Division of Reactor Development and Technology, has stated that "It becomes more evident each day how dependent we are going to become on the successful introduction of breeders in order to be assured of practically limitless economic electric power and process heat."[62] So far, however, it is clear that economic and practicable breeders lie some way in the future, despite intensive research. Our own view is that a serious query about the realistic feasibility of breeders persists. In any case, we may discount breeders for present purposes in view of their merely prospective status.

In passing, we may observe that nuclear power has little, if any, relevance to the so-called underdeveloped countries (i.e. two-thirds of the world), since the "... cost of modernisation and industrialization required to utilize the electric power exceeds the cost of the power itself by several orders of magnitude."[63]

We may now turn to consideration of some of the limitations of, and objections to, nuclear energy. First, there is the question of raw materials. Here we are concerned principally with uranium-235.

Nuclear burners use up large quantities of uranium-235. But this material is in short supply, since it has an abundance of only 0.7% of the uranium in natural ore.[64] It has been calculated that: "If reactor development proceeds as foreseen by the Atomic Energy Commission, inexpense reserves of uranium (costing less than $10 per pound) would be used up within about 15 years and medium-priced fuel (up to $30 per pound) would be used up by the year 2000."[65] Since that projection was made, however, "... the estimate of nuclear power plant capacity for 1980 has been increased from 95,000 to 145,000 electrical megawatts without a corresponding increase in the estimates of uranium reserves."[66] The same source continues: "An even further restriction arises from the rate at which these reserves can be mined and processed. According to Faulkner, of the reasonably assured reserves of 310,000 tons of U_3O_8 in the United States, only about 210,000 tons can be produced by 1980. His corresponding estimate for cumulative world production is about 500,000 tons. This alone could force the low-priced reserves into a higher-price category in case, as appears likely, it should be necessary to double the rate of production."[67]

The survey of raw materials would be incomplete unless we pointed out that, as regards the hoped-for fusion reactor, there is an important limitation in the scarcity of lithium, an essential element in the lithium-deuterium fusion reaction.[67a]

There is, therefore, it is obvious, a desperate scientific race against time involved in the matter of nuclear energy. Can the transition to breeders be accomplished before the initial supply of uranium-235 is exhausted? We do not know, of course, but we remain sceptical. We would, however, agree with the scientific verdict that "Failure to make this transition would constitute one of the major disasters in human history."[68] This judgement is based upon a careful appraisal of fossil fuel prospects, and assumes that these alone cannot sustain human development (for want of a better term) very much longer.

We may proceed to attempt an assessment of nuclear prospects. First, we may note that, just as with early steam engines, the nature of atomic engines is such that where mobility and flexibility are important required characteristics they have little relevance. In other words, it seems unlikely that atomic engines will ever replace the internal combustion engine (or its electrical successor) for the purposes these serve.[68a]

It has been said that "Taking a view of not less than a century, were electrical power to continue to be produced solely by the present type of light water (i.e. burner) reactors, the entire episode of nuclear energy would probably be short lived ... With the use nuclear power would no longer be economically competitive with that from fuels and water."[69] Thinking ahead to the prospect (which seems to us remote) of breeder reactors the judgement has been passed that: "Even the breeder reactor will give us neither free nor unlimited power. The cost of nuclear fuel for the breeder reactor will indeed be negligible; but the cost of the large capital investment, of power transmission, of waste disposal, and of operation combine to bring the likely best price per kilowatt hour to about that of cheap steam-coal electric generating plants. Breeder reactors will be a wonderful asset to industrial nations not because they provide cheaper power but because they may provide desperately needed power when the fossil fuels are depleted and before the fusion reactor can be perfected."[70] It is also worthy of comment that, given the perfection of techniques for extracting the few parts per million of uranium and thorium in granite, we do not have the slightest idea – far less plan – about what to do with the leftover granite, in a world already critically afflicted with dereliction.[71]

We should surmise, on the basis of the available information, that nuclear power *may* meet about 10% of American energy requirements by the year 2000. At the moment, however, of 65 American nuclear facilities due to be operational by 1976, 23 are behind schedule.[73] Our surmise may, therefore, veer on the optimistic side. Either way, nuclear energy "... will not be a major source of energy in this decade".[74] The Assistant Secretary for Economic Affairs of the American state Department was well briefed when he conceded at the end of 1970 that the fossil fuels dominate the energy field and will continue to do so.[75] As for Britain, one who is on the whole a nuclear optimist has cautiously restricted himself to saying that "... nuclear power stations will prove a *useful part-insurance* for Britain against possible economic or political obstructions in the importation of oil and in the production or importation of coal."[76] (emphasis added)

The questions we have hitherto been tackling are, however, only part of the story. We must accordingly now turn to the important matters of nuclear pollution and the disposal of nuclear waste. In our opinion it may be these considerations rather than the more technical and economic sketched above that determine the ultimate contribution of nuclear energy to the global energy equation. It is to these matters that we now accordingly turn.

With pollution we may couple the risk of accident. A number of books have recently drawn attention to the dangers inherent in the nuclear power programme.[77] One of the authors has summed up the problem succinctly in these words: "... the environment which supports us has only a limited capacity for radiation, and that capacity can only be used once."[78] Isotopes escape into the air and water all the time from nuclear plants. It appears that for the foreseeable future there is no satisfactory way of stopping this continuous inadvertent emission. There have already been cases where the concentration of escaped isotopes has constituted a threat to contiguous populations. "The projected release of one isotope *alone,* krypton-85 could, claims Ehrlich, "within the next century raise the level of radiation exposure of the general population to 60% of the maximum permissible level set by the National Committee on Radiation Protection and Measurement. Krypton-85 is just one of some 200 radioactive substances released at 'low levels' by fission reactors."[79] The history of the last 25 years teaches us, moreover, that official bodies time and again err dangerously on the side of optimism in establishing "maximum permissible levels."[80] Scientists of repute have time and again appealed for quite new and more stringent safety guidelines, for otherwise the whole human race is threatened with biological deterioration. We can do no better than reproduce here, with permission, a recent article in the authoritative anti-nuclear war paper *Sanity* which summarises what is at stake.[80a] The article follows:

"If America's safety levels against radiation from peaceful atomic energy installations are not substantially lowered, the country may suffer: 90,000 additional deaths from cancer yearly; 60,000 more pre-natal deaths each year; 2,000 extra cases of leukaemia; and 12,000 additional children will be born with gross mental and physical defects. This is an estimate published by Prof. Linus Pauling, Nobel prizewinner and Professor of Chemistry at Stanford University in the *Bulletin of the Atomic Scientists.*

Urgent warnings have been published also by two other scientists, Dr. A.R. Tamplin of the Lawrence Radiation Laboratory, USA, and Dr. J.W. Gofman, Professor of Medical Physics at the University of California, who estimate 32,000 additional cancer deaths a year.

These statements have, of course, brought a strong counter-attack from America's powerful Atomic Energy Commission and from leaders of the nuclear industry, who deny the scientists' findings.

The warnings have now aroused the interest - and fears - of much wider sections of the population, and, says the *Bulletin,* 'The controversy has evolved into a national concern.'

Calculations made by Dr. Tamplin and Dr. Gofman follow estimates by Dr. Ernest Sternglass (published in *Sanity* last year) that fallout from nuclear weapons tests has been responsible for 400,000 infant deaths. They support warnings uttered 10 years ago by Prof. Pauling in his book *No More War!* which protested against the biological effects of nuclear weapons tests.

Tamplin and Gofman, also writing in the *Bulletin* say: 'It must come as a shock to the lay public that approximately 25 years into the Atomic Era, we should be in the midst of a raging controversy concerning the biological effects of radiation. There are two important reasons why this controversy has surfaced at this time. One is the burgeoning nuclear industry associated with the peaceful uses of nuclear energy. The other is the sudden and dramatic increase in the nuclear weapons project created by the development of the MIRV and the ABM programmes. One of the most essential inputs to the nuclear projects for both peace and war is the biological effects of radiation. It is important to point out that the present controversy is merely a resurfacing of Linus Pauling's earlier estimate of the biological effects of radiation. The major difference today is that the data accumulated over the past 10 years demonstrated that Pauling's original estimate was substantially correct. A second difference is that the Atomic Energy Commission and the Congressional Joint Committee on Atomic Energy have cleverly manipulated the nuclear power industry to join them as bedfellows in this controversy. The position of all three is indefensible and arrogant.'

The allowable dose of radiation set by the US Federal Radiation Council is at present 170 millirad a year, but Tamplin and Gofman say the dose should immediately be reduced to 17 millirad, otherwise thousands of people will die needlessly each year.

In his Nobel Peace Prize lecture on December 10, 1963, Linus Pauling said that because of radioactive pollution caused by nuclear bomb tests 'about two million people now living will die five or ten or 15 years earlier than if the nuclear tests had not been made.'

Tamplin and Gofman comment: 'The data which have accumulated since his Nobel lecture in 1963 demonstrate that Pauling's opinion was correct. The data demonstrate that all forms of cancer can be induced by radiation. Moreover, they show that the various cancers are induced in proportion to their normal occurrence rate. As a consequence, the present data indicate that the effects of a given dosage of radiation is ten times worse than it was thought to be in 1963.'

Linus Pauling in his *Bulletin* article says of his estimate of 90,000 additional deaths yearly from cancer caused by radiation from peaceful nuclear installations: 'This estimate is larger than that of Gofman and Tamplin, who calculated that there would be about 30,000 additional cases of leukemia per year in the United States if everyone received the Federal Radiation Council's statutory allowable doses of high energy radiation. Gofman and Tamplin state that their estimate, for several reasons, which they give, is to be considered a minimum. My estimate is neither a minimum nor a maximum; it is the estimate that seems to me to be indicated as the most probable by the available evidence.'

Prof. Pauling concludes: 'We may ask whether the sacrifice of some tens of thousands of people to save the money that would have to be spent to decrease the amount of exposure to high energy radiation is justified. People die ultimately; if not from cancer then from some other disease. Need we be concerned that some people are caused, by our decision, to die five or ten or fifteen years earlier than they would have died if our decision had been a different one? I feel that we should be concerned; that the cutting off of a man's life in this way by cancer is undesirable; that we should try to decrease the number of deaths by cancer rather than take such action as to increase their number. Also, much suffering is caused by the birth of a grossly defective child; I believe we should strive to decrease the number of such births.'

The AEC, replying to Dr. Tamplin and Dr. Gofman in the *Bulletin,* agrees that all radiation is potentially dangerous and that radiation exposure should always be kept as low as possible. But it disputes the calculations of the two scientists and their conclusions. It says: '... the Gofman-Tamplin position assumes that every person in the United States somehow received 170 millirads a year from the nuclear power programme. That cannot physically occur or even be remotely approached.'"

This article, in our opinion, makes very clear what is at stake in the nuclear power programme. If safety measures were taken to satisfy the criteria of Pauling, Gofman and Tamplin nuclear power plants could not possibly compete, in terms of cost, with conventional fossil fuel plants. They would quite simply be priced out of consideration. These facts will, of course, be kept as far as possible from the awareness of the average citizen by interested parties such as governments and nuclear contractors. Ultimately, however, they cannot be suppressed.

For a time it was euphorically thought that the advent of fusion plants would remove or mitigate the pollution danger. Quite apart from the fact that controlled fusion reaction remains a totally theoretical possibility, this hope in any case now appears unfounded. F.L. Parker of the International Atomic Energy Agency contends that "... the escape of radioactive tritium from fusion power plants may prove even more hazardous than the escape of isotopes from fission reactors."[81]

Another pollution danger inevitably associated with nuclear power is thermal pollution. The release of hot industrial wastes into streams and lakes "... is an extremely grave threat to acquatic life, much of which is highly sensitive to temperature change. Nuclear power plants in particular are serious thermal polluters. On the average they waste 60% more energy than plants that burn fossil fuels. It has been estimated that by 1980 nuclear plants alone will be using 20% of the total fresh water runoff of the United States for cooling."[82] Little imagination is required to grasp the magnitude of the ecological disaster implied in this statistic.

The pollution danger is only one aspect of the matter. Another is the possibility of accidents. It is not sufficiently understood how terribly vulnerable present nuclear plants are. Private insurance companies refuse to cover the risks. Were it not for massive public subsidies, the existing nuclear power establishments could not continue.[83] Ehrlich writes: "The reluctance of private companies to supply nuclear power plants with liability insurance is based in part on the 'near misses' in the reactor field, such as the accident in 1966 at the Fermi Plant outside Detroit, which potentially could have killed millions of people and rendered a substantial part of the US uninhabitable. Prior to 1964 there had been twelve reactor accidents involving serious damage to the installation, radiation overexposure for individuals, or release of radioactivity to the environment. Some of these incidents exceeded the 'maximum credible accident' for the installation involved. This concept in itself is an indication of the euphemistic AEC approach to safety, for the 'maximum credible accident' is defined as the worst one which would occur in the absence of human error and with all safety devices working perfectly. Since human beings make mistakes and safety devices are prone to failure, this is hardly reassuring. It is clear that until the AEC can be reorganised to provide cautious and intelligent control of immediate and long-term enviornmental hazards, constant vigilance by Congress and concerned citizens will be necessary to avoid running grave risks."[84] It is to be feared that this liberal perspective on the powers of "Congress and concerned citizens" is wildly optimistic in view of the evidence that exists of the superior power of organised business and governmental vested interest groups.

How long human luck will hold before there is a catastrophic nuclear power plant accident is a matter of conjecture. Recently, however, Ralph Lapp, a prominent nuclear physicist, writing in the magazine *New Republic,* forecast that a serious nuclear accident would appear to be a certainty before the year 2000. He argued that a nuclear reactor " ... constitutes a unique hazard to people and property in its vicinity."[85] He continued: "New criteria emerge, more unknowns are identified and more research is indicated, but all the while more powerful reactors are being constructed closer to cities." In his considered view, the basic safety issue is whether "... full reliance can or should be placed on the inherent safety of the reactors and their engineered safety features ... (or in) ... minimising risk through the inter-position of distance between the reactor and the population." Nothing, he maintains, exists to substantiate public confidence in pronouncements of the Atomic Energy Commission that the benefits to be gained from nuclear reactors "far outnumber the risks of the potential hazard."

A major nuclear power plant accident would certainly be a disaster without strict precedent. But it is not the only risk we run by pursuing an energy policy increasingly dependent upon nuclear generation.

There is also a pressing waste disposal problem. Nuclear waste is not like other waste. There is no way of hastening its reduction to a harmless state.[86] All that we can attempt is to isolate it as effectively as possible while it is lethal.

The human body has accommodated itself to what is known as background radiation – that is, the naturally occuring level of radiation emitted by our environment. Any increase in this level, however, is likely to induce harmful biological changes in animals, including humans, exposed to it. The process of power production by nuclear fission is such that "... the mass of the radioactive fission products produced in a reactor is very nearly equal to the mass of fuel consumed."[87] The question is, basically what to do with this stuff for the 600 to 1,000 years it may take to decay to the point where it is biologically harmless? As more and more nuclear installations are constructed the problem becomes more and more urgent.

The AEC commissioned an expert advisory committee on waste disposal in 1955. It contained a variety of experts in such different relevant fields as geology, hydrology and mining. The committee came up with the following three guidelines for disposal:

"1. All radioactive materials are biologically injurious. Therefore, radioactive wastes should be isolated from the biological environment during their periods of harmfulness, which for long-lived isotopes exceeds 600 years.

2. The rate of generation of radioactive waste is roughly proportional to the rate of power production from nuclear fission reactors. In the period of its work, the committee regarded the rate of nuclear power and related radioactive waste production as being on the very low portion of a steep exponential-growth curve. The committee therefore reasoned that no waste-disposal practice, even if regarded as safe at an initially low level of waste production, should be initiated unless it would still be safe when the rate of waste production becomes orders of magnitude greater.

3. No compromise of safety in the interest of economy of waste disposal should be tolerated."[89]

Looking at existing practices in these terms, it has been said that most – other than those concerned with high-level liquid – " ... violate the first of the three principles stated above, and probably the second also."[89a] In other words, they are not being successfully isolated from the biological environment, and they are of dubious applicability when the rate of waste production increases ten or a hundred times over.

We have had sufficient experience now to have learned empirically that pumping vast amounts of waste deep into the ground through convenient or induced faults or direct into underground reservoirs is an intrinsically dangerous procedure. Not only is it virtually impossible to guarantee that water which will eventually re-enter surface circulation will not be affected. It is also now evident that there may be dangerous geological repercussions previously unforeseen: "In 1967 the consequences of four years of pumping fluid chemical wastes into an underground reservoir near Denver became clear. A series of earthquakes occurred, the three largest of which had magnitudes of about 5; slight damage was reported in Denver. The amount of energy released in the series of earthquakes was slightly greater than that released by a one kiloton A-bomb, more energy than was expended in pumping the fluid into the reservoir. The remaining energy had been stored in the Earth's crust by geologic processes, and its release was triggered by the injection of fluid into the underground reservoir."[90] In this particular instance, the result was not catastrophic. If such practices continue, however, it appears likely that – sooner or later – there will be a geological accident of disaster proportions.[91]

There remains the question of transportation of the wastes to disposal locations: "It has been estimated that by the year 2000, more than 3000 6-ton trucks will be in transit at any given time carrying such wastes to burial sites. Truck accidents will be a constant serious threat."[92] There have already, of course, been mishaps. Fortunately, up to now, these have been minor or quickly brought under control. How long will it be at present rates before a more serious accident occurs? The following report seems likely to become familiar reading in the decades ahead:

"Radioactive Scare: Sydney, Tues. - All traffic was diverted of (sic) a busy express-way here early this morning when radioactive material was discovered on the roadway. Police said the radioactive material was believed to have fallen from a truck. They did not know what the material was. - Reuter '"[93] A minor incident, to be sure, but an ominous portent for the future.

We may conclude this consideration of the dangers inextricably associated with the development of nuclear power by pointing out that if it proves impossible to remove them by ingenuity, conscious planning, and much expenditure – and we do not hold out much (if any) hope for this –

the entire nuclear programme may prove, for this reason alone, to have been a false starter.[94]

Notes

In what follows, the following abbreviations have been observed:

ARU = A.R. Ubbelohde; *Man and Energy*, London, 1963.
Ehrlich = Paul R. Ehrlich and Anne H. Ehrlich: *Population, Resources, Environment*, San Francisco, 1970.
NAS = National Academy of Sciences-National Research Council Committee on Resources and Man: *Resources and Man*, San Francisco, 1969.
SA = *Scientific American*, September 1970 (a special issue on the Biosphere).
All other references are fully cited.

59. Ehrlich, p.57.
60. NAS, pp.219 et seq.
61. NAS, p.221.
62. NAS, p.223.
63. Ehrlich, p.5 7.
64. SA, p.187.
65. SA, p.187.
66. NAS, p.224.
67. NAS, p.224.
67a. NAS, p.233; Professor Nicol raises another pertinent question about nuclear energy and inputs: "The cost of nuclear energy in terms of conventional fuels is difficult to compute; it is certainly no bargain, being far more than the chemical equivalent of processed uranium or other nuclear fuel." *(The Limits of Man,* P.134).
68. NAS, p.228.
68a. ARU, pp.54-55.
69. NAS, p.226.
70. NAS, p.122.
71. Ehrlich, p.57; see also John Barr; *Derelict Britain,* London, 1970.
72. SA. p.184.
73. *The Economist,* London, 23/1/71.
74. *The Economist,* London, 23/1/71.
75. Philip H. Tresize in *Department of State Bulletin,* 26/10/70.
76. ARU, p. 79.
77. See, for example, R. Curtis and Elizabeth Hogan: *Perils of the Peaceful Atom,* New York, 1969; Sheldon Novick: *The Careless Atom* Boston 1969.
78. Sheldon Novick: *The Careless Atom,* p.

79. Ehrlich, p.13 7.
80. See Ehrlich, p.138.
80a. We are grateful to Philip Bolsover, the Editor of *Sanity* for permission to reproduce the article here. The issue was that for February, 1971..
81. Ehrlich, pp.137-138.
82. Ehrlich, p.18 7; "The efficiency of a power plant is determined by the laws of thermodynamics. No matter what the fuel is, one tries to create high temperature steam for driving the turbines and to condense the steam at the least possible temperature. Water is the only possible medium for carrying the heat away. Hence more than 80% of the cooling water used by U.S. industry is accounted for by electric power plants. For every kilowatt-hour of energy produced about 6,000 B.T.U. in heat must be dissipated from a fossil fuel plant and about 10,000 B.T.U. from a contemporary nuclear plant. In the U.S. where the consumption of power has been doubling every eight to 10 years, the increase in the number and size of electric power plant is putting severe strain on the supply of cooling water. By 1980 about half of the normal runoff of fresh water will be needed for this purpose. Even though some 95% of the water thus used is returned to the stream, it is not the same; its increased temperature has a number of harmful effects. Higher temperatures decrease the amount of dissolved oxygen and therefore the capacity of the stream to assimilate organic wastes. Bacterial decomposition is accelerated, further depressing the oxygen level. The reduction of oxygen decreases the viability of acquatic organisms while at the same time the higher temperature raises their metabolic rate and therefore their need for oxygen." (SA, p.190).
83. Ehrlich, p.138.
84. Ehrlich, p.139.
85. Reported in *The Straits Times,* Singapore, 22/1 /7 l.
86. "Each radioactive isotope decays at a fixed negative-exponential rate peculiar to itself" (NAS, p.233).
87. NAS, p.234.
88. NAS, pp.234-235
89. NAS, p.236.
89a. NAS, p.236.
90. Ehrlich, p.141.
91. See story headed "Earthquake fears over atom test" in *The Straits Times,*10/5/71; see also transcript of Thames TV programme "And on the Eighth Day". 27 /l/70.
92. Eblich, p.137.
93. *The Straits Times,* Singapore, 5/5/71.
94. Soothing bromides notwithstanding; see: M.J. Gaines: *Atomic Energy,* London, 1969, pp. 112-114. Mr. Gaines is a science writer with the U.K. Atomic Energy Authority.

The Politics of Nuclear Energy

Alan Roberts

Alan Roberts (1943-1990) joined the Labour Party in 1959 and CND the following year. Elected to Parliament at the 1979 General Election, Roberts supported Tony Benn's bid to be elected Deputy Leader, opposed the Falklands War and carried out important work on housing. His other Spokesman publication, "Consumerism" and the Ecological Crisis, is available in facsimile.

This article was first published in Hazards of Nuclear Power *(Spokesman, 1977) together with Zhores Medvedev's article 'Nuclear Disasters in the Soviet Union'.*

The political importance of nuclear power

Modern capitalism has turned increasingly towards technological "advances" that are suspect in the extreme. They are marked by their dubious or plainly negative contribution to human welfare, and by their destructive effects on the environment.

There are some whose harmfulness is now widely recognised – as, for example, the replacement of efficient public transport by a commitment to the private car, the switch to detergents, the massive use of pesticides, the waste of energy in packaging (particularly the non-returnable bottle and the aluminium can).[1]

It is now clear, however, that one particular development – the nuclear power industry – looms above all others, in its ominous implications for the future of humanity, and in its significance as an issue on which mass action against the system's irrationality is likely.

Its predominance derives, firstly, from the sheer magnitude of the economic commitment involved. The leading capitalist countries intend to generate most of their electrical power by nuclear means before the turn of the century, necessitating an unprecedented speed of construction. Over the next decade alone, the US government hopes to see nuclear capacity increased eight-fold; France and Japan aim at roughly fifteen-fold growth. These programmes imply that the USA, for instance, is to spend well over a trillion (million million) dollars on the nuclear industry in the next two and a half decades.[2] It has been estimated that, if the 1985 target is achieved, the nuclear power industry will absorb *over fifty per cent* of gross US capital formation over the next decade.

Next in importance is the transparency of the irrationality involved. It is not a matter of waiting till consequences difficult to foresee have come to pass – as, for example, it was necessary for the polluting effect of detergents actually to show themselves, or for the cities to become congested, polluted and deformed by the automobile. The damage inherent in the nuclear development can be clearly foreseen at this very moment.

The third feature is one of special significance for social change; it concerns the response of the populations in the advanced capitalist countries once they are reached by the arguments against nuclear power. Outstanding here is the example of Sweden, the only country where the issue has been made the subject of more or less formal nation-wide discussion. These discussions, carried on in the course of the year 1974, saw the population swing from approval of the nuclear programme to better than two-to-one opposition. As a result, the government cut its ten-year nuclear target to one-seventh of its former size (from fourteen reactors to two).[3]

Similar responses on a more local scale have been evident in the USA, where the nuclear industry openly expresses its fear that nuclear moratoria (federal or state) will be imposed as a result of public opposition.[4]

Thus it is not simply a question of a valid issue, implying a struggle for all concerned with humanity's future. The campaign against the nuclear commitment also has the character of a transitional demand, striking at the very assumptions of consumerist society, and yet understandable to and acceptable by the people affected.

In countries of the Third World, the political context of the nuclear issue is different but the validity of the struggle is no less clear. It is necessary to emphasise this point particularly, since the proponents of nuclear power often advance arguments allegedly based on the interests of a power starved Third World – arguments which, as we will see, could hardly be more specious.

Why the nuclear programmes are unacceptable

The dangers associated with nuclear power have been adequately explained in a number of publications, and here we will simply refer the reader to them.[5] They fall under the following main headings:

1. Unscheduled discharges of radiation to the environment, in amounts exceeding the low levels prescribed in normal operation.
2. Catastrophic releases of fuel or waste materials, following on a "meltdown" of the fuel after an accident.
3. Deliberate release (or the threat of it), of radioactive materials, as a measure of terrorism or criminal extortion.
4. Environmental damage arising from nuclear wastes (whose disposal remains an unsolved problem).
5. Undesirable political and social measures adopted to cope with these hazards.

The possible magnitude of some of these dangers can be judged from the simple facts concerning the highly toxic element plutonium. The maximum permissible annual intake of plutonium is at present one millionth of a gram, a quantity known to be capable of causing cancer (and considered too high a risk by many authorities, including Britain's Medical Research Council).[6] But the most common type of nuclear reactor, in normal operation, over one year, produces about 200 kilograms of plutonium.

Of course, stringent precautions are taken to ensure that this and other radioactive poisons are contained and never reach the atmosphere. But no system of containment can be perfect, nor verified with absolute accuracy. (Today, for example, the inventory of plutonium in a reactor cannot be checked to better than 1%.)

Suppose then that, by the end of the century, when upwards of 2,000 reactors are envisaged, a small fraction of the plutonium generated in a year "leaks" to the atmosphere – whether by accident or malevolent design. If the leak is as small as one hundredth of one per cent of the total, this still constitutes a maximum permissible dose for every person in the world, *ten times over.*

The nuclear programme thus embodies a proposal to organise power production around stocks of highly poisonous substances, in quantities almost unimaginably vast in relation to their toxicity. To accept such a programme, one would need to be supremely confident of the social system in which it is to be implemented – confident both of its ability to maintain unprecedentedly high standards of technical skill with absolutely infallible rigour, and of its political and social stability over many generations. The reader can be presumed to lack such confidence.

Despite the quite extraordinary and often ingenious safety routines implemented by the nuclear technologists, whose efforts to achieve the impossible must compel admiration, the safety of the US nuclear industry has already been the target of damaging criticisms. These concern the workings of about fifty reactors in the world's most industrially advanced country; what can be expected when perhaps 2,000 reactors are operating in dozens of countries throughout the world?

Some indication of an answer to this question was given by Jean-Claude Leny, managing director of Francatome. It took the form of a broad hint to investors, that the profitability of nuclear power in France would not be allowed to suffer – like the American industry's – from an exaggerated concern for safety ...[7]

As for the possibility of malevolent activity, the infant nuclear industry of the USA can already record, amongst other incidents, a threat to crash a highjacked plane into a reactor, a series of apparent sabotage attempts in a re-processing plant, and the selection of nuclear plants for terrorist blackmail attacks by followers of Charles Manson.[8]

It should be remembered that the possible damage arising from nuclear catastrophes is not confined to the existing population in the country of occurrence.

The very nature of the radioactive threat lends itself to dispersal in space over national and even continental boundaries, and to persistence in time so that generations remote from the present suffer illness and death (the genetic effects of radiation). The lesson from the USA in particular is that the industry's safety standards will tend to be proportional to public concern over the issue; in this light, the struggle against nuclear power can be seen also as a simple struggle for human survival on the planet.

The disposal of waste materials from reactors – and of the worn-out reactors themselves – remains an unsolved problem. Its magnitude can be gauged from one figure alone: the annual wastes from an average reactor today contain 1,000 times the radioactivity of the Hiroshima bomb. While research proceeds on possible methods of permanent disposal, the industry contents itself with "waste management" – that is, retrievable and (it is hoped) secure methods of storage. Here it should be noted that the cost of this "temporary" storage (which is by no means at a satisfactory level of security) will rise in the next two and a half decades to some seven billion dollars in the United States alone. It is easy, then, to understand the fear expressed by US Environmental Protection Agency experts, of "the possibility that an interim engineered storage system may become permanent solely due to economic costs".[9]

To understand the ominous implications here, one should first note that the interim methods make the poisonous waste "retrievable" – or in other words, accessible. Thus they continually invite malevolently-inspired acquisition or atmospheric release. Also, the time scale of the "permanent" storage required is not in dispute: the long-lasting component of the wastes (particularly plutonium) must be kept rigorously clear of the environment for hundreds of thousands of years – half a million, for safety. This poses the unprecedented problem of finding a storage which will not be disturbed by the *geological* processes that occur over such a time span. Research has not yet proved that such storage exists. Here, once again, an issue of sheer survival is involved, in the struggle to prevent such irresponsibility towards future generations.

The nuclear industry has generally treated critics with disdain, making concessions to them reluctantly and only after public opinion has been roused. But in recent years, some of the more far-sighted proponents of nuclear power have started to recognize the strength of the opposition's case, particularly in the area of "nuclear malevolence". Their proposals for coping with nuclear hazards constitute in themselves an equally ominous political and social threat.

Thus the US Atomic Energy Commission has proposed a special federal police force devoted to the security of plutonium plants and shipments. It has complained of recent court rulings protecting individual privacy, and requested legislation which would facilitate security checks on nuclear industry personnel.[10]

With the projected growth of the industry, the number of workers affected by such restrictions of civil rights could run into the millions. Already, according to the *New York Times,* Texas state police keep dossiers on opponents of nuclear plants.[11]

The dangers involved here should not be underestimated. A few kilograms of plutonium make an ideal weapon for blackmailing a whole city, since it effectively disperses itself in small particles once exposed to the air. Even graver is the real possibility of constructing a nuclear bomb from plutonium in a reactor's waste, impurities would make it inefficient but, as an experiment has convincingly shown, little skill would be needed to achieve a weapon with the destructive force of about 100 tons of TNT.[12] This would be within the capacity of "amateurs", any government with nuclear power plants would have the facilities to manufacture weapons 100 times more deadly.

After an extortion threat, whether successful or not, an atmosphere of hysteria could well be envisaged, in which authoritarian "law and order" proposals would be difficult to combat. They would even have a certain rationality, inside a globally irrational context.

The many levels of irrationality

The risks just outlined justify the verdict that a major development of nuclear power is irrational, if our criterion is the welfare of humanity. But this is far from the only sense in which we can justly apply the epithet "irrational" to capitalism's nuclear perspectives.

It should first be appreciated that the current nuclear programme is not a long-term solution to the problems of power generation, even in the opinion of capitalism's own analysts. It is projected as merely bridging the gap between the present period marked by rapidly diminishing stocks of oil, and the situation in perhaps three decades or so, when alternative sources of energy will be commercially viable.

The tapping of the sun's energy is one important such alternative, to which capitalism is now belatedly starting to devote increased research and development funds. The primary aim here is to find ways of reducing the capital costs of large-scale solar power plants.

For reasons discussed below, solar power is still seen as less attractive than fusion power – a variety of nuclear plant working on a different principle from the current models. Existing "fission" reactors rely on a controlled version of the nuclear reaction – the "splitting" of a heavy atom such as uranium or plutonium – which in its convulsive release produced the explosion of the Hiroshima bomb. A "fusion" reactor would be based on taming the nuclear reaction underlying the hydrogen bomb, in which light elements "fuse" together to form a heavier element. Steady progress is being made in the research on controlled fusion, particularly since a Soviet breakthrough in this field some years ago – the "Tokamak" development. It is generally believed, however, that several decades will elapse before commercial fusion reactors enter into service, even after a basic design has proved itself in the laboratory.

Thus present nuclear programmes are supposed to justify themselves by their contribution to power needs in the next few decades. But it is precisely in this short term that there arise the most serious doubts of the programme's utility, because of the severe shortage of rich uranium ores.

The industry's major hope here lies in the breeder reactor, whose operating core is wrapped in a "blanket" of natural uranium. Such a reactor will convert the bulk of this uranium into a suitable fuel (normally, less than one per cent of it is available), thus producing (or "breeding") more fuel than it uses up. The world supplies of "burnable" uranium could thus be effectively increased perhaps 70 times over.[13,14]

Before agreeing with the US administration that breeder reactors thus represent the solution to the nuclear fuel shortage, some facts should be noted. The inherent dangers of the breeder reactor vastly exceed those of the current models, and justify the greater concern and opposition of aware scientists.[15] A whole series of technical difficulties have resulted in repeated postponements of the expected date of operation of a commercial breeder, the latest estimate (probably optimistic) now landing in the 1990s.

The significantly higher capital costs, as compared to today's power stations, are likely to result in yet more delays before the buying reluctance of electrical utilities is overcome. And even then, a breeder will take somewhere between 20 and 40 years to produce enough fuel for one reactor.

Thus, reliance on the advent of breeders to "stretch" fuel supplies represents a dubious gamble.

Yet what the industry is thereby gambling on, is the whole cost-competitiveness of nuclear power.

It is irrationality of another sort which emerges here: the nuclear programme is not even rational on capitalism's own criterion of cost efficiency. Reactors already planned are not assured of a fuel supply which can keep them competitive. Thousands of billions of dollars are to be invested in the hope that something will turn up.

Even with the cheap uranium supply available today, the industry can establish the competitiveness of new plants only by ignoring well-established trends, that would send the price of nuclear-generated electricity skyrocketing. The most important of these trends are, firstly, the staggering escalation in the capital cost of nuclear plants, and secondly, the severe drop in efficiency of nuclear plants after about five years' running.

In May 1975, the Friends of the Earth showed how woefully the relevant utility had underestimated costs, when they testified against the proposed Rancho Seco 2 reactor near Sacramento (California). Adopting realistic figures for capital cost, interest rates and capacity factor (i.e. efficiency), and for operation, maintenance and decommissioning, the FOE calculation showed that the *true* cost of a unit of power was nearly *four times* the figure submitted by the utility.[16]

A study of the Grenoble Institute has shown that, in France, nuclear-generated electricity cannot compete with oil at today's prices. In the heating of a household, for example, we can deduce from the study that oil will be cheaper so long as its price remains below $45 a barrel (price in early 1977: approximately $16).[17]

The escalation in capital cost (we consider its explanation later) shows no sign of abating. Of course, that of coal-fired plants also shows an increasing trend, but nothing like as severe – a 1975 study estimated that the *difference* in price between a coal and a nuclear plant was itself increasing by $19 per kilowatt per year.[18] In other words; every year the price of a 1000-megawatt nuclear plant leaps another $19 million above that of its coal-burning rival ...

The curves of capacity factor against reactor age also show a dismal trend: that the efficiency is low and becoming even lower.[19] All this may make the nuclear commitment seem extraordinary enough; but we have not yet mentioned the most astonishing irrationality of all. Some preliminary remarks are needed.

The power output of a generator of any sort can never represent pure gain, since some power is inevitably consumed in building and running it. In the case of a nuclear reactor, a great deal of power is required merely to set it up in business – to build the station, mine and mill the initial fuel supply, etc. A most important part of this power input occurs at the stage where natural uranium is treated so as to increase the fraction of it which can be "burnt" as fuel – the "enrichment" process.

All this means that the station will be running for some time before it has "paid back" the power used to set it into operation. Calculations of this "break-even" time have been carried out for various reactor designs; they indicate that about two years of normal operation will be needed to repay the power input for construction.

Now consider the effect of a rapid nuclear *programme,* with the number of reactors doubling every few years. To see this effect, let us adopt some definite (though fictitious) figures: suppose a reactor's "pay-back" time is one year (this is unrealistically low), and that the number of reactors is doubling every year (this is unrealistically fast). Suppose also that a reactor takes a year to build (instead of the actual six to nine years).

In year one, no reactors are operating but one is being built; so no power is produced, but one year's output is consumed. In year two, one reactor is operating, but two are under construction, so one year's output is produced, but two are consumed. In year three, three reactors are operating but four are being built; so three years' output is produced, but four are consumed ...

If the calculation is continued it will be found that the programme uses up more power than it produces, *in every year of its operation.* Of course, in the real world such a programme must come to a halt at some stage, the number of reactors cannot go on doubling each year indefinitely. It is at this point that the nuclear industry will become a net power producer; but until then, it will actually be a net *consumer* of power.

In the real world, also, the figures are not as they are given in the example. But the effect still persists in a modified form, even after we insert the correct data for power input in operation and building time. We still find that the programme will not "break even", in the sense of producing more power than it consumers, for a certain number of years.

Just how many years, will depend on a number of factors: the type of reactor, its operating efficiency, the grade of ore mined, the power consumed in regular operation. But the most detailed calculations available[20] suggest that, inserting the figures appropriate to current programmes, this "break-even" time can easily exceed 20 years.

But this is precisely the period in which the nuclear programme is supposed to compensate for the exhaustion of oil supplies, while the world awaits the arrival of fresh power sources. In other words, the nuclear programme will quite possibly consume more power than it produces, in the very period when it is supposed to be the key factor in power generation!

It should be pointed out that a programme with oil- and coal-burning stations[14] substituted for nuclear, but expanding just as quickly, would make an even worse showing. It is the sheer *speed* of the projected construction programmes which determines their short-term energy inefficiency. But of course, no one plans to build conventional power stations at such a breakneck pace – since no one has the illusion that such a programme would solve any "energy crisis". This illusion attaches only to plans for nuclear power stations, when one "forgets" the energy needed to build them; to puncture the illusion, the sort of energy analysis sketched above is required.

Before arriving at an overall judgement on capitalism's nuclear project, we should appreciate the element of uncertainty which runs through the above analyses. Some of the needed data – what fresh reserves of uranium will be discovered, for instance, or what long-term efficiency (capacity factor) will be achieved by nuclear stations – can only be estimated. Some of the relevant calculations require time and manpower that have not yet been devoted to them, so that only suggestive approximations are available.

However, this very absence of reliable information is itself highly revealing. Let us adopt some of the criteria commonly advanced, *within* a framework of capitalist assumptions, for implementing a new technology, and consider how they are met in the case of nuclear power. Let us see what preconditions should be fulfilled to justify the investment of capital involved.

First, the safety of the new industry should be sufficiently guaranteed, as to obviate the risk of the whole development being aborted at some future date. (This could occur, for example, as a sequel to the catastrophic release of radioactive material, by a plant accident or malevolent design. The public reaction could well make it politically impossible to continue operation of the existing plants, and force the abandonment of the large amounts of capital they represented.)

Secondly, the programmes adopted should actually achieve their declared goals: that is, to produce significantly more power than they consume, in the vital period of the next few decades.

Thirdly, the electricity produced should be competitive in cost with that generated by "conventional"(oil- or coal-fired) stations.

Fourthly, plants should not be projected unless they are guaranteed a suitable supply of fuel over their working lifetime.

Fifthly, the financial mechanisms should exist that will enable the "consumer" (i.e. the electrical utilities) to obtain the capital needed to buy the reactors concerned.

It is when we review these reasonable criteria that there emerges the full irrationality of capitalism's nuclear plans: it has not been demonstrated that they satisfy a *single one* of these basic requirements.

At best, the nuclear industrialists can be regarded as undertaking a colossal gamble. They are gambling that no catastrophic accident will occur in the short term, despite the narrow squeaks already in the record. They are gambling that fresh high-grade ore reserves, or a technically and commercially viable breeder reactor, will be available in time. They are gambling that the trend to ever-higher capital costs, and the decline with age in the efficiency of the functioning reactors, will be reversed, or economically compensated for by increased cost of conventional fuels.

In the USA, they are even gambling that "something will turn up" in the way of finance, to permit the purchase of reactors by the electrical utilities. (Early in 1975, some 603 of reactor orders in the USA had been cancelled or postponed, mainly because of the refusal of finance houses to lend the purchase money.)[21]

It is true that capitalist enterprises have been known to "gamble" before this – to spend on research and development, or to launch on the production of a new commodity whose market was not assured. But we remind the reader of the sums involved in this particular gamble – well over a thousand billion dollars in the remainder of this century, in the United States alone.

It would be easy to conclude that the gods of history, with the destruction of capitalism high on their agenda, are staging their proverbial prologue of induced lunacy. But a pat verdict of "guilty but insane", even if supported by the evidence, hardly goes far enough; it is also necessary to *understand*.

The attempt to reach even a partial understanding is mandatory, and not only because of the importance of the nuclear programme in itself, both economically and politically. There is another issue involved: that of the dynamic of the capitalist economy in the present period. It may be that the nuclear industry can serve as a paradigm, showing – in not-so-small miniature – the emergence of new trends or changes in the relative weight of ones already known.

The Energy Company's Gamble

There are few industries, even today, as heavily monopolised as the nuclear industry. When one says "pressurised-water reactor", one says Westinghouse; and "boiling-water reactor" likewise means General Electric. And these two types, built by the two giants directly or through subsidiaries and licensing agents throughout the capitalist world, account for over 85% of the nuclear-component industry.

The powerful pressure of these multi-national corporations exerts itself even on those countries possessing their own proven reactor designs. Thus Francis Perrin, formerly the French high commissioner for atomic energy, has recently complained of the "monolithism" of the French nuclear programme (even while rubbishing the anti-nuclear campaign as "based only on totally false assertions" and on declarations "devoid of all objective value").

He recalls General de Gaulle's decision (December 12, 1967) to proceed with the construction of two large reactors of a French design (graphite-moderated, gas-cooled, fuelled by natural uranium) that has elsewhere proved itself. The blocking of this decision he lays to the account only of some unnamed highly-placed civil servants, also responsible for the present plan to install "almost exclusively" the pressurised-water reactors of ... Westinghouse.

He calls, but without much apparent faith in the likelihood of success, for the French programme to include more "diversification", a feature not sufficiently provided by the present inclusion of some boiling-water reactors from ... General Electric.[22]

The weight of the multi-nationals has been felt even in Britain, the country whose own design of gas-cooled reactor pioneered the commercial generation of nuclear electricity. Hot debate raged after the Central Electricity Generating Board and the National Nuclear Corporation both recommended a switch to the American light-water reactor. But under intensive questioning before a House of Commons Select Committee, they were unable to justify their recommendations, and the Government decided not to switch – for the time being, at least.

The revelations from Lockheed and other firms have made notorious one of the processes by which the multi-nationals "conquer" foreign markets. It should not be assumed, however, that this is always the predominant factor. Sheer size counts for a great deal – as illustrated in the unhappy case of the design of an international computing language. The world's experts agreed on a suitable language, and devoted much effort to its elaboration. But their eugenic offspring, Algol, runs a very poor second in its breadth of social acceptance to the inferior language, Fortran – which was born with a silver spoon in its mouth, sired by the market-dominating IBM.

In another direction, a still vaster oligopolistic structure is shaping up, as the leading oil companies complete their transformation into what has been accurately described as "energy companies". Already in 1971, the oil giants were responsible for the milling of some 40% of US uranium; their coal production amounted to 20% of the US total, and their acquisition of coal reserves guaranteed their future dominance in the industry (one oil company alone – Humble – was the nation's second largest coal owner). In the nuclear field, Gulf Oil (with the third largest assets – about $9 billion – of any oil company) had set up Gulf General Atomic.[23]

This latter company threatens Britain's lead in gas-cooled reactors, and already in 1972 there was "consternation in the nuclear industry" as a consequence, according to one writer.[24] Gulf promises delivery of high-temperature gas reactors (an advanced design) around 1980.

But if this represents competition with the dominant light-water American reactors, is similar consternation apparent among the ruling giants? Hardly; the chairman of Gulf General Atomic, E. Prockett, happens to sit on the board of Westinghouse also.

A thrust towards monopolisation is built into the nuclear project. A single plant of today's typical size – a thousand megawatts of electrical power – costs upwards of half a billion dollars, and smaller units are neither readily available nor called for in quantity. Companies with assets not running into the billions can hardly hope for a sizeable share of such a market, nor risk the investments needed to establish themselves.

The dynamic of capitalism's nuclear project has been spelled out – with some naive admiration – by Simon Rippon, the editor of a technical journal noted for its fervent, not to say fanatical, nuclear partisanship.

"...The big industrial concerns have not entered the business for quick profits - indeed, most of the companies that have entered the nuclear business around the world have been shaken to their foundations by losses on early projects and few can see dramatic profits in the future. For the most part the position of industry is that the long term direction of energy supply is going to be increasingly in the direction of nuclear power and therefore for the well-being of their company they must establish a foothold in this sector of the business in spite of the heavy initial costs."[25]

It may be doubted whether the "foothold" is being seized as reluctantly as Rippon makes it sound. For the larger giants, nuclear power spells centralisation, size, growth. The prospect before them is an intoxicating one: the power industry swollen to a size unheard of, its relative weight in the economy enhanced several times over, and all of it within the grasp of one or two amicably coexisting combines.

The power industry as a whole can of course anticipate such an increase in its relative share of the gross national product, since the power needs of industrial capitalist society grow faster than the GNP itself. In Japan, for instance, official projections are for a growth of 4% in the GNP, compared to 6.2% for the electrical output.[26] Using this data, a simple calculation shows that the proportion of the GNP represented by electricity output (i.e. its relative weight in the economy) will be double what it is now, in a little over 30 years.

It is only this perspective which can explain the gambles they are taking, and pressuring governments to take. They are not really gambling that no catastrophes will occur, that no hitches will hold up the breeder reactor when it is needed, that the nuclear project will remain cost-competitive.

What they are really gambling on – and from their viewpoint, it is a "rational" risk to take – is that their economic and especially their political weight in society will be so massive, that society has no option but to make their bets come home.

It is the next decade which is crucial for this outcome. By 1985, the nuclear share in electricity production is designed to reach, in the leading capitalist countries, the 10% level or close to it (the USA, 13%; the EEC, 17%; France, 30%).

Within the present structure of industrial capitalism, it is hard to envisage a situation in which such proportions of the power supply could simply be switched off, no matter how powerful the arguments in terms of human welfare or even of economic efficiency.

Perhaps a catastrophic "melt-down", releasing millions of curies of radioactivity, killing tens of thousands of people, damaging property to the extent of billions of dollars? Studies by the American Atomic Energy Commission have shown that accidents could well have such a scope.[27] But if society really depends on the nuclear branch of its power industry in order to continue along its accustomed path, and if this path can still claim an overall acceptance, then an alternative to a shutdown would be the adoption of "firm measures", allegedly ensuring that such disasters could not recur.

Such measures, whose shape was sketched in the AEC report mentioned earlier, would be repressive and authoritarian in the extreme; and there can be little doubt that among the movements heavily repressed would be any spreading panic or mobilizing action in connection with nuclear power.

But if nuclear power reveals itself as unarguably wasteful? Suppose the tendencies for nuclear plants to decline in efficiency with age, and to require more and more capital for their construction, become so pronounced that, on economic grounds, they should simply be replaced by non-nuclear methods of power generation. Would not this be a situation disastrous to the nuclear industry, one in which their gamble had definitively failed? Possibly – if they allowed such a situation to arise. But, as a Harvard-MIT study pointed out in the *Technology Review:*

> "The price of usable energy from oil, coal or uranium now has little to do with the marginal production cost of any of these resources ... Instead, the price of energy from alternative technologies is the result of a complicated process of assigning relative values to a variety of energy-producing resources and technologies by those who either control or require these resources and technologies. This process is both intensely and inherently political."[18]

In assessing the degree of control over energy prices, it is vital to realize that we are not dealing with an isolated handful of reactor manufacturers – more and more, the Energy Company becomes a powerful reality, and the relative pricing of the various methods of electricity generation falls increasingly under its control. "Free competition" between the various primary fuels started to lose its reality many years ago, as the oil companies moved over into the mining of coal, of uranium, into the processing of uraniums and – through subsidiaries and affiliates – into the building of reactors. Their influence will be exerted to fix prices that reflect, not the resultant of competitive forces, and not the realities of cost of effectiveness but simply the interests of their own needs for expansion, investment and profit.

Thus, if the nuclear industry is gambling, it knows in advance that the dice will be loaded in its favour. And even if its luck turns unexpectedly bad, and the table runs against it incessantly, there remains a further and decisive recourse: it can have a word with the management ...

Consumerist capitalism needs the power industry; it even needs its continuous and sizeable expansion. The State which administers that system never runs on the basis of one-capitalist-one-vote, or even one million-dollars-one-vote; always some animals in that particular jungle play the role of the king of beasts. The Energy Company, more than half nuclearised by the turn of the century, will certainly supply a king or two, perhaps even a king of kings. Such personages do not need to fear bankruptcy, or even a missed dividend. If even the smaller predators like Lockheed, Boeing or Gruman can depend on sympathetic intervention by the State in their hour of need, what will be beyond the power of the Energy Company?

Indeed, nuclear power has already benefited crucially from State support, and not only in the billions lavished on research and development, whose results the corporations simply take over. Another important parcel of "aid" has been delivered by the US government plants enriching uranium. The Westinghouse and GE reactors require fuel that has passed through this expensive process, and their success in penetrating the market is due in no small measure to the artificially-low price assured by what amounts to a concealed State subsidy; an advantage which has not gone unnoticed by their competitors:

> "Ned Franklin, chairman and managing director of Britain's Nuclear Power Company ... maintains that the price of uranium enrichment is now fixed by essentially political considerations. Enrichment is dominated by the US, which supplies most of the enrichment requirements of the western world. According to people working in the US's nuclear industry, the prevailing price of enrichment is about half what it would be if the industry had to build new facilities and operate them at a profit.
>
> The problem is that enrichment is subsidised by the use of old plant that was paid for as part of the weapons programme; enrichment plants are supplied with subsidized electricity; and there is no charge for research and development."[28]

With such marks of favour already acquired, there seems little that the Energy Company needs to fear – unless, of course, it confronts an enemy whom even the State must treat with caution.

Creating the "objective facts"

The socialist movement has suffered for many generations from the illusion that technology is value-free. Adopting a misleading schema in which an essentially non-political "base" (the forces of production) is to be simply taken over and endowed with a different "superstructure" (socialist relations of production), it has failed to appreciate the political content of that technological base.

Even Lenin is on record as succumbing to this error, when he went so far as to laud the Taylor system (time and motion study) and urge its adoption in the Soviet Union. It should be noted that a question mark must now be put over the "technological rationality" of the assembly-line method itself; can it really be justified even on the narrow criterion of "stepping up production"? This most alienating of all technological practices needs re-examination in the light of recent industrial experiments (particularly in Sweden) based on a self-managed working team, rather than a single worker permanently assigned to one stultifying operation on the line.

That technology, and the line of development of technology, are alike political, is nowhere more evident today than in capitalism's nuclear project. It is illuminating to consider the non-nuclear alternatives for power supply, their undesirability from monopoly capital's viewpoint, and the way that an apparently inevitable technological progress along nuclear lines is actually the result of highly political decisions.

A source of nuclear power has supplied mankind with the overwhelming bulk of its energy throughout history; it is the sun, a giant reactor successfully employing the fusion process without pollution and without wasting non-renewable fuel reserves (over a time scale of several billions of years, at any rate). Serious studies of the world's energy problems almost invariably urge the priority of research and development in the field of solar power as the most attractive prospect for mankind.

But it might be asked: how real is this prospect of solar power? What are the technological data on its practicability as a large-scale resource? How does its level of development compare with other energy sources, and what is its promise in the short term?

Questions such as these are posed at the wrong level; they seek as answers a recital of "bare" technological data, not themselves embodying politico-economic decisions, but supplying the value-free facts on which such decisions can be based. It is true that there are circumstances (very restricted, and usually of little social interest) in which such a dichotomy of fact and value has a relative validity; but the present questions are not located in a context even remotely appropriate to such a division.

Large-scale nuclear reactors actually exist; nuclear power moved out of the laboratory many decades ago, into the province of the architect and the engineer. Large-scale solar plants, on the contrary, remain in the anteroom of research and development. Is this a "bare" technological fact? Only in the most abstract sense; in the real world, the genesis, understanding and future implications of this "fact" must be sought in the sphere of political economy.

For there is no autonomous, independently evolving sphere of "technological progress" which thus made nuclear plants arrive before solar. Nuclear technology was developed in response to conscious decisions on the allocation of manpower and funds – inspired originally by the search for more destructive weapons, and later by the attractiveness for monopoly capitalism of the peculiar qualities of nuclear power.

The failure to allocate corresponding resources to solar power research was the complementary decision that helped to create the "technological facts" as they now exist. And of course, similar remarks can be made about projects to tap the earth's subterranean heat (geothermal power), or to utilize the tides.

Thus the facts are purely technological only in abstraction, inside a conceptual schema that isolates from its social context an abstract history of "technological progress". In the concrete world of things as they have been and as they are, these facts are born already "dressed" in a political-economic penumbra that accompanies them always, determines their significance and points to their future possibilities.

This can be seen very clearly, when we consider the prospects of solar power *vis-a-vis* nuclear, over the next couple of decades. The "facts" involved here are being created right now, and a glance at US budgetary allocations will show us what facts the Energy Company hopes to bring about: for every dollar spent this year on solar research, more than eight dollars will be spent on one nuclear project alone – the breeder reactor.[29]

It is not hard to understand why monopoly capital is so lukewarm towards solar power. The latter lends itself admirably to decentralisation, small installations, a minimum investment of capital; these are fatal flaws from the viewpoint of the giant corporation. The "technical" advantages – inexhaustible energy supply, absence of pollution, longevity of the installation, low maintenance expenses – cannot compensate for these inbuilt deficiencies ... It has been well said, that solar power would fare very differently if only General Electric could buy the sun!

The sad fact is, however, that solar leases are not yet open to takeover bids; and so the corporations are doing the next best thing: planning to build their own sun ... For there *is* some corporation interest in solar power, provided the inbuilt vices just mentioned can be eliminated, and the project made capital-intensive, large-scale, highly centralised. These are precisely the qualities of the Satellite Solar Power Station, emanating from Arthur D. Little Inc., Grumman, Raytheon and Textron. A giant satellite a kilometre across will absorb sunlight, convert it to microwave radiation and beam it down to a seven-kilometre receiver on the Earth's surface, generating from three to 15 times the output of a single large nuclear plant.[30]

In principle, the solar power source can be a highly flexible device, adaptable in size to meet a wide range of demand and providing access to power for the most isolated community. A minimum of capital investment can provide a self-sufficient source for an indefinite period, and one uniquely compatible with ecological requirements.

These features can hardly be recognized in the satellite project, which achieves the near-impossible: a solar power source demanding an enormous capital investment, suitable for insertion into only the very largest national electricity grids, taking no advantage of solar radiation's great suitability for direct heating of homes and workplaces, and delivering, with its giant receiving antennae, an insult to the environment on a new and monstrous scale.

We do monopoly capital an injustice, then, if we evaluate its nuclear programme as nothing more than a technological project. Quite apart from its inherent hazards to humanity, its adoption would then become incomprehensible in view of the serious doubts as to nett energy production, security of investment, reliability of fuel supply and cost-competitiveness.

But actually it must be seen as a project in a much wider sense: namely, as a *social* project, predicated upon a definite social structure and aiming to develop that structure in a definite direction.

The social structure concerned is that of capitalism in its consumerist phase, where a widening gap – between a potential for self-managing fulfilment, and a reality of hierarchical repression – is papered over with a policy of consumerist concessions. Destruction of the environment is implicit in such a society; this connection has been analysed in some detail elsewhere, and will not be further discussed here.[31]

The power needs of such a society are vast and ever increasing, and it indeed faces a 'crisis' in the prospect of exhaustion of oil reserves, combined with a severe pollution problem from coal-burning power sources. But, for reasons which will be clear from the discussion above, the giant corporations which dominate its technical development can hardly be enthusiastic about the rational lines of solution advocated even by its own experts: elimination of wasteful energy consumption, reduction in the growth of the electrical power industry, development of alternative sources such as solar, geothermal and tidal power.

It is true that nuclear power, too, has its disadvantages – it may, for example, weaken the fabric of social control by the destructive or blackmailing opportunities it creates for dissident groups. But in lending itself to centralisation, expansion, and domination by a few industrial giants, it accords well with the dynamic of consumerist capitalism – which would be hard put to accommodate policies of energy conservation and the strangling of growth.

Of course, the system will have to adjust itself to the peculiarities of this new power source. The Energy Company may have to distort market and pricing mechanisms more grotesquely still, to nudge along the consumption of nuclear-generated electricity and the purchase of nuclear reactors. Massive and direct State intervention may be required to ensure the industry's future, with the perhaps grudging consent, or even against the opposition, of industrialists in other sectors. And measures of social discipline will almost certainly be called for, restricting civil rights and limiting the activities of protest movements, to provide the safeguards needed once society depends for its life-blood – electrical power – on one or two thousand incredibly poisonous sources. Such expectations may well appear repugnant, but they cannot be dubbed fantastic; they are solidly based on existing values and assumptions, those which demand the constant expansion of the commodity market and, to an even greater extent, of electricity output.

But these values and assumptions do not go unchallenged, and there is nothing fatalistically inevitable about the scenario sketched above. We have been looking at the political economy of capitalism today; but a different political economy is also shaping itself, already in conflict with its older rival and by no means invariably vanquished. We must now look at the forces behind this alternative view, take note of their accomplishments up to the present and estimate their possibilities in the future.

The Political Economy of Contestation

Opposition to the construction of nuclear power plants has developed, over the last five years, into a world-wide campaign of significant scope and impact. Despite the power of the corporative forces committed to the nuclear programme, the journals of the nuclear industry overtly and repeatedly express the fears roused in them by the achievements and potential of their opponents. "Things can't get worse or can they?' was the gloomy title of an editorial in *Nuclear News* (April 1975), which went on:

> "The likelihood of a nuclear moratorium, either national or in one or more states, is difficult to assess. Judged from the discussion of it among observers of the Washington DC scene, and from the amount of activity on the state level, the situation is not encouraging for the light-water reactor industry, and is much worse for the breeder reactor."

A writer in *Nuclear Engineering International* (July 1974, p.579) raised a similar possibility: "However unjustified, public opposition to nuclear energy may rise to such levels that forecast installation programmes have to be scrapped ...".

Superficially, some of the nuclear industry's major troubles seem unconnected with the anti-nuclear opposition. We have seen how, in early 1975, about 60 per cent of the nuclear plants on order had been deferred or cancelled – a severe blow to the Administration's nuclear plans forming part of "Project Independence". This setback is usually attributed to the "cash squeeze"' of the time, which made Wall Street reluctant to lend the electrical utilities the capital with which to purchase reactors.

It is true that some orders for "conventional" power stations were likewise affected; but even so, the finance houses do not seem too enthusiastic about the economic future of nuclear-generated electricity. Nor are they alone in their doubts.

Robert F. Gilkeson, chairman of the Edison Electric Institute, was reported as saying at the April 1975 American Power Conference that "it is impossible in present circumstances to build a power plant that will yield a satisfactory return on investment."[32] After analysing the poor performance of the older reactors, David Corney doubts if the banking community will be willing to finance the nuclear programme, and suggests that General Electric, Westinghouse and other nuclear firms may "join Lockheed, Boeing and Grumman on the rolls of corporations bailed out of costly technological misadventure by the taxpayers."

It might seem that here, at any rate, we have unearthed some "bare" technological facts which, despite all their contortions and figure-juggling, the nuclear corporations cannot conceal. Nuclear power is just too costly, and that's that ... Or is it? Let us investigate a little more deeply:

Nuclear power stations are usually situated well away from the densely populated areas in which the electricity is actually consumed. This entails a two-fold economic penalty, as Hohenemser points out:

First, that part of the energy released which is not converted into electricity becomes pure waste, since the consumers are not sufficiently near to allow this energy to be used for residential and commercial heating and cooling. Thus the very promising concept of a "total energy system" cannot be realised, and the surplus energy becomes waste heat whose disposal is a problem. But the energy thus wasted is more than double the electrical energy utilized.

Secondly, the additional distance over which electricity must be transmitted means additional investment in transmission lines, and additional losses in energy.

Furthermore, conservative operating procedures are adopted to prevent possible accidents; operating costs rise because of the need to protect workers from radiation. As Hohenemser sums it up: "The accident risk, though small, leads to large economic penalties."

It will be apparent that these economic penalties cannot be regarded as solely economic in origin. The pressures which force the nuclear station to be sited remotely, or to adopt stringent and costly precautions, depend intimately on the level of popular suspicion of nuclear power, and of legal-political activity based upon that suspicion.

Thus it is difficult to interpret these economic difficulties of nuclear power as pure "technological data". But further analysis makes the point emerge even more sharply:

Perhaps the most important single factor telling against the economic future of nuclear power is the continuing escalation in capital cost of the nuclear plants, as compared to coal-burning plants. The reasons for this escalation have been carefully analysed in *Technology Review* (February 1975) by Bupp (Harvard) and Derian, Donsimoni and Treitel (MIT).

They find that total cost is strongly correlated with the length of the licensing period – i.e. the time elapsed before the plant is licensed by the Atomic Energy Commission (AEC) to enter into operation. Under US law, citizens can "intervene", on safety, environmental and other grounds, to oppose the granting of the licence or secure its postponement.

It is this intervention process, they show, which carries the responsibility for prolongation of the licensing period and the correlated rise in capital costs:

"The American administrative and judicial processes afford ... critics ample opportunity to impede the rate of reactor commercialisation. The principal consequence has been dramatic cost increases. The extreme critics of nuclear power have been at least partially successful in their efforts to force a downward re-evaluation of the social value of reactor technology.

... The issue here is not merely technical or economical, but is inherently political: Present trends in nuclear reactor costs can be interpreted as the economic result of a fundamental debate on nuclear power within the US community. Beyond its economic effects, the real issue of this debate is the social acceptability of nuclear power ..."

(It should perhaps be recalled that critics of nuclear power are not free to hold up construction at will; they must show that the particular project fails to satisfy environmental requirements, existing radiation-release standards, AEC regulations ... And it is precisely this kind of deficiency that they have been able to establish, time and again.)

Perhaps the second most ominous trend, for nuclear-power competitiveness, is that of declining capacity factor (efficiency) as plants grow older. A detailed study of the reasons for this decline is still in progress, but some contributing factors are already apparent, which are associated with the radioactive dangers in a nuclear plant and the public consciousness of them. For instance, the discovery in September 1974 of cracks in the cooling pipes of a US reactor resulted in the shutting-down (for inspection) of all reactors of the same type; this would hardly have been done in the case of conventional power stations. Nor would it have been done, in all probability, if the public were less inclined to associate danger with the word "nuclear".

Unprecedented maintenance difficulties can arise in nuclear reactors; the simple welding of a crack becomes a large-scale operation in which hundreds of workers have to be deployed, when the crack occurs in a region of such high radioactivity that each worker can remain there for no longer than a few minutes ... Here again, the long campaign which forced the AEC to tighten up its radiation standards, and the heightened public awareness which resulted, should not be overlooked as a relevant factor.

We see, then, that the Energy Company has not got the field to itself; there are other political choices and actions which are significantly affecting the "bare economic facts" of nuclear power production. And of course, their effect on the political decisions in this field is even more noticeable – as shown, for example, by the severe reduction in the Swedish nuclear programme for the next decade (from 14 reactors to two) already mentioned above.

We will not go on to list the successes of the anti-nuclear campaign in such other countries as Japan; the above is enough to show that significant effects can be achieved. This is all the more remarkable, being given that most of the radical left, in most of these struggles, have followed a policy of more-or less benevolent abstention.

It should be said, in conclusion, that the anti-nuclear movement is likely to find its path much thornier in the future. The year 1975 must be recorded as the year of the great backlash, when the nuclear industry geared itself up on an international scale to launch a well-organised counter offensive.

In Washington, a pro-nuclear rally was scheduled for the middle of May – "The first time that the industry, which has traditionally avoided direct action on its own behalf, has set out to make itself heard", according to a supporter. This rally was to unite representatives of the Atomic Industrial Forum, the non-profit utilities, the National Association of Electric Companies (investor-owned utilities) and the national rural electrical co-operative association.[33]

In April, the European Nuclear Society met in Paris, at a conference reported as though it were a similar propagandist rally.[34] Westinghouse assigned a team of propagandists in Pittsburgh to the job of "rebutting" environmentalist objections to nuclear power stations.[35] The Atomic Energy Commission in Australia – a country with no commercial reactors – ran an internal study course for its staff, slanted towards the justification of nuclear power. (The export of uranium is a current issue in Australia.)

In launching this propaganda offensive on a global scale, the corporations tacitly acknowledge both the importance of the nuclear development for the immediate future of consumerist capitalism, and their appreciation of the strength of mass suspicion in its regard. It is vital that the left show an equal appreciation of these factors, participating wholeheartedly in the anti-nuclear campaign and strengthening its connection with the overall struggle against an irrational social system.

The left is hampered in fulfilling this role by the misleading theory (among others discussed further on) that the technological sphere evolves autonomously, independent of political action. The philosophical defects in this view have been surveyed above; after considering the particular case of the nuclear power industry, we can see how woefully it fails to explain the facts and the dynamic of this major component of capitalist planning in the decades to come.

Of course, the traditional marxist view never entirely overlooked this phenomenon; but it was usually content with a mere mention of the existence of "reciprocal interaction", or of the "mutual independence" of the various sectors of the social "totality". The analysis itself usually proceeded in a strictly one-way direction, with the political exercising little if any direct influence on the technological or economic.

It would be wrong to claim that this method has now lost all validity; but it is apparent that, in the case of nuclear power, it does not give even a good first approximation to the truth. It is difficult to conceive of this holding good only for one special and exceptional case, when that case looms so large in terms of economic significance and investment allocation. Are we not rather looking at a paradigm of capitalism's development in this present phase, with deep lessons for the left and its programme of radical reconstruction?

Whatever the misconceptions of some of its practitioners, marxism could never have been properly interpreted as a variety of economic determinism, in which technological development exerted a one-way influence on the remaining structures of society. Marxism separated itself decisively from such theories by its standpoint of class analysis, so that the technological sphere can be effective only when mediated through the prevailing class interests.

The interests of the capitalist class are not to be conceived as simply the making of a fast buck. They include also the preservation of a structure of industry which will enable the capitalist system to continue; and it is precisely this continuance of the centralised, large-scale, ever-expanding economy, based on a market of "created demand", which the environmental crises today put in serious doubt.

In this situation, the larger investment decisions must be seen as political decisions, in which the longer-term interests of the system must take precedence over narrowly-conceived "economic" interests. But as political acts, they become vulnerable to the attacks of political opponents – a vulnerability which the outstandingly irrational nuclear industry knows only too well, as it nurses its wounds and lashes back.

Thus, in intervening in struggles over the shape of the economy the left should not be hampered by any lingering compunctions, perhaps based on recollections of the "Luddite" period, of the "utopian machine-wreckers" (recollections which are revealed as obsolete by the facts above, and which were generally inaccurate historically in any case). Otherwise, they will be leaving unchallenged some of the most significant political decisions of the giant corporations, carrying immediate threats to the world of today and even sowing the seeds of disaster for humanity's whole future.

A Digression: the USSR and the "Third World"

The analysis above is focused on the advanced capitalist countries, and should not be extrapolated beyond them. The other major sectors of the world merit a separate if briefer discussion.

With a total list of only 25 plants, including those under construction or on order, the nuclear programme of the USSR is insignificant in comparison to that of the USA, which is some 15 times greater in power output. Indeed, France's alone outstrips the Soviet's in capacity (by about 50 per cent).[36]

This lesser level of development is not to be explained by an initial technological lag – the first Soviet nuclear station opened in 1958, ahead of every other country in the world save one (Britain).

Nor does it stem from any ideological aversion to nuclear power. Official Soviet doctrine sees no problem in the inherent centralized nature of nuclear power; no problem in the superhuman standards demanded for safe operation in the long term: no problem in the disposal of radioactive wastes.

Indeed, the absence of genuine public discussion on the issues involved in nuclear power has allowed the Soviet nuclear industry to "solve" its disposal problems with a breathtaking light mindedness: high-level radioactive wastes are simply pumped under pressure into deep permeable zones. Thus they are irretrievable; in insecure liquid form; and moreover (because of the high pressure of the injection), a threat to the stability of the whole region; disposal methods with these objectionable features would never be permitted in the USA or Europe.[37]

In explaining the Soviet tardiness in nuclear development, one cannot overlook the abundance of its coal, oil and hydropower resources. But the absence of private ownership also seems relevant here, saving the USSR from some of the more spectacularly irrational features of capitalism's technological policies. At least its power supply will not be shaped by the imperial adventures of an Energy Company.

The situation of nuclear power in the Third World is of direct relevance to the controversy in the industrially advanced capitalist countries. For defenders of nuclear power there often rest their case on the needs of Third World countries; short of coal, faced with rising oil prices, and yet starved of energy for their economic take-off, their only hope, allegedly, is the power of the atom.

This argument is either cynical or simply ignorant. A United Nations analysis has revealed the true situation, referring first to the Third World's

"... very poor infrastructure of technology and non-availability of trained manpower to handle the reactors and other nuclear plants. The probability of nuclear accidents and consequently of dangers to human environment are bound to be far greater in these countries. Further it is doubtful whether these countries could afford to spend an additional $3-4 billion towards the foreign exchange cost of nuclear facilities during the next 25 years which will be the years of financial stress in these countries arising from pressure of population and scarcity of food. Moreover, the small size of the national electric power grids can integrate only small nuclear power plants which are at present not being manufactured ..."[38]

This last point is at present vital: the leading corporations are simply not interested in building reactors small enough to fit Third World needs. And they appear to remain adamant despite pleas by nuclear protagonists in the specialist literature, and even by leading figures at the September 1974 conference of the International Atomic Energy Authority.[39]

Evidently they prefer to fight one battle at a time. Once the developed "heartland" has been conquered for nuclear power, it may be time to think of the outskirts.

The people of the Third World have no interest in speeding up the process of their "nuclearisation"; the UN comments above show this clearly enough. Financially, the higher capital cost of nuclear plants would deepen their dependence on the imperialist countries, who are skilled in exacting a political price for "development loans". Technologically, an important part of their industry would be in the hands of metropolitan experts for several decades. Economically, even a medium-sized plant would usually constitute by itself a high degree of concentration of power supply, and favour a centralization of industry and a grandiosity of construction squarely opposed to the real needs of the bulk of the population. (When the majority of the population have no access to a power point, the arrival of a nuclear plant can hardly do otherwise than distort the economy further. What benefits have flowed through to the mass of people in those underdeveloped countries already boasting nuclear stations – Pakistan, India, Spain?)

The Role of the Left

In the campaign against nuclear power – as in most of the campaigns on environmental issues – it has been exceptional to find the political vanguards actually in the van ... In its most extreme form, this suspicion leads to a dismissal of the anti-nuclear struggle – indeed of environmentalist issues in general – as a trendy middle-class phenomenon that does not interest the working class, and hence is no concern of the true revolutionary, who will concentrate on the real issues: those at the point of production and in the realm of State power.

Such a class characterisation of the environmentalist movement has greater difficulty reconciling itself with the facts now, than it might have had a few years ago; a weakness more serious still, is the implied judgement of an issue, not on its merits as a valid transitional demand, but on its present level of working-class penetration.

It might be worth pointing out how neatly this attitude reverses the approach to social problems that was typical of Karl Marx. Absorbed above all else by humanity's need for the overthrow of capitalism, Marx had an eagle eye – whether as journalist or as theoretician – for movements which contained the seed of revolution. Seeing the revolutionary potential of the working class, he thereafter focused his theoretical and practical activity on the needs and development of the working class movement.

The attitude we are examining turns this upside down. An attachment to the role of the working class – or rather, to a particular selection from Marx's writings about it in his day – serves it as a reason for ignoring what was Marx's first concern: evidence of revolutionary potential in any movements or strata in the contemporary world. If such schools of thought turn a blind eye to the environmental movement, their vision is not much keener when it comes to the liberation movements of women, blacks or gays. Eventually, after the passage of time, some Galileo may be able to persuade them to look through his telescope. But they will need first to be convinced that the sights they will see can somehow (perhaps tortuously) be reconciled with the true reality – which for them (as it never was for Marx) is constituted by their *doctrine.*

A widespread climate of such opinions can exert a damaging influence – as it appears to have done even to a talented and perceptive analyst such as Hans-Magnus Enzenberger. His article, 'A Critique of Political Ecology', dissects and exposes some of the best-publicised "doomsday ecologists", such as Ehrlich, in a study of considerable value. But the reader will search in vain for any recommendation that the left should participate in, and endeavour to guide, mass movements to defend the environment – from nuclear contamination or anything much else.[40]

Despite Enzenberger's clear recognition of the possibility of what he calls "ecological rebellions" and "uncontrollable riots", he is uneasy about the "dangers" of participation by the left, and can only recommend that "a long process of clarification will be necessary..."

By confining itself to the study and to a role of instruction from afar, the left will indeed avoid the risk of being "used" – just as an army is in no danger of being tricked and outmanoeuvred if it keeps clear of the battlefield. But, specialising from environmental issues in general to the nuclear question in particular, it must be asked whether the ground should really be surrendered to the enemy so easily.

The historical import of the nuclear power programme derives from the current plight of modern capitalism: based firmly on consumerist values and concessions, it sees the development of that consumerism heading inexorably towards the destruction of the environment. The coming exhaustion of oil reserves is one harbinger of the crisis, and has prompted a reckless acceleration of the nuclear programmes, in an attempt to ensure, at whatever cost, that consumerist capitalism will have available the centralised sources of power it needs.

The struggle over nuclear power thus poses questions about the very shape of society itself – as any intervention in this struggle quickly reveals. For it is impossible to adopt a purely negative stance, attacking nuclear power but proposing no alternative energy policy.

Many of the reformist critics understand this well, and offer programmes which envisage the attainment of social energy goals without the use of nuclear power, but which usually involve sizeable reductions in energy consumption by various methods of conservation.

But such a conservation policy would represent an extraordinary historical "turn" by a consumerist capitalist society, wedded as it is to continual expansion; a society, moreover, in which the relative weight of the "Energy Company" grows day by day. Can such a society significantly restrict its energy consumption over a whole business cycle – for example, in a time of recession, will it throttle down on vitally needed expansion plans, simply because they are energy-expensive? And what would be the social and political reverberations of such energy-conserving policies as were adopted?

These important questions usually get scant consideration from moderate advocates of conservation. In contrast, those already convinced of the need for radical social change are less inhibited, and will not play down the severe strains which an energy crisis implies for capitalism today. But their own social project will not escape a similar critique, unless it has at least the basic outline of a solution to the problem – unless it can point to the satisfactions it envisages as replacing the dubious rewards of the commodity culture.

One project which sketches such a solution is that of self-managed socialism. The substitution of the principle of self-management for the present dominant principle of hierarchy in every walk of life – a substitution possible only if the power of the capitalist is overthrown and that of the bureaucrat severely limited at least – implies on the level of the individual, the possibility of changing the values one lives by. If new channels of self-expression and autonomous action can be opened up in every social sphere, beginning with the factory floor, it will not be so crushing a catastrophe if beer must be brought in bottles rather than in energy-expensive aluminium cans.

This point has been made in greater detail elsewhere.[41] It illustrates how the campaign against nuclear power must be finally unconvincing, unless it is prepared to delineate an alternative social path, a credible one that does not lead to a poisoned world. A receptive atmosphere for such an exposition is created by the striking irrationality of the nuclear programme, which must condemn by association the system that gives rise to it, and encourage the consideration of rational alternatives.

Notes

1. Discussed in *The Closing Circle,* by Barry Commoner, Jonathan Cape 1972.
2. "Nuclear Electric Power", by David J. Rose, page 359. *Science, 184* 19 April 1974.
3. *Nuclear News,* April 1975, p.80. (The ruling Swedish Social Democrats subsequently lost the general election in which their remaining plans for nuclear plants were severely challenged.)
4. *Nuclear News,* April 1975, p.33 (editorial).
5. The best source here is *Non-nuclear futures,* by Amory B. Lovins and John H. Price (Ballinger Publishing
Company, Cambridge Mass.), October 1975, which contains an encyclopaedic list of references.
6. *Nature, 253,* p.385 (February 6, 1975, editorial).
7. *Investir,* March 24 1975. Quoted in Basquet, *Le Nouvel Observateur,* April 21 1975, p.46.
8. For the latter two incidents, see respectively *Environment,16,* October 1974, page 21, and *Time,* September 22 1975 ("Fromme: 'There is a Gun Pointed'").
9. "The Hidden Commitment of Nuclear Wastes", by W.D. Rowe and W.F. Holcomb. *Nuclear Technology,24,* December 1974, p.286.
10. "Plutonium Recycle: The Fateful Step", by J.G. Speth, A.R. Tamplin and T.B. Cochran. *Bulletin of the Atomic Scientists,* November 1974, page 19.
11. *Ibid.,* page 20.
12. *New Scientist,* March 27 1975, page 799.
13. A 1000 Megawatt (electrical) reactor requires about 4,500 tons of uranium over its lifetime. Thus a world total of 2,000 reactors (one of the lower estimates) by 2000 AD would need some 9 million tons; but the estimated world inventory extractable at less than $39 a kilogram is 4 million tons. (See e.g. "World Uranium Resources", by L.G. Poole, *Nuclear Engineering International,* February 1975).
14. See the discussion in Speth *et.al* (ref. 10 above).

15. See e.g. "A Troublesome Brew", by Sheldon Novick, and "A Poor Buy", by T.B. Cochran *et.al*, both in *Environment, 17,June* 1975.
16. "The Deflation of Rancho Seco 2", by Jim Harding. Reprinted from *Not Man Apart* (undated).
17. *Alternatives au nucleaire,* Presses universitaires de Grenoble, February 1975. The figure cited follows from Annexe 3, page 89, on utilising the findings on capital cost from ref.18 below, and those on capacity factor from ref.19 below.
18. "The Economics of Nuclear Power", by LC. Bupp and J.C. Derian *et.al. Technology Review,* February 1975.
19. "Will Idle Capacity Kill Nuclear Power?" by D.D. Corney. *Bulletin of the Atomic Scientists,* November 1974.
20. In Part 2 ("Dynamic Energy Analysis and Nuclear Power") of *Non-nuclear Futures,* by A.B. Lovins and J.H. Price. Ballinger (Cambridge Mass.), 1975.
21. *Nuclear Engineering International,* February 1975, page 73.
22. *Le Nouvel Observateur,* April 28, 1975, page 86.
23. "The World Energy Market", by B.C. Netschert, *Bulletin of the Atomic Scientists,* October 19 71.
24. *New Scientist,* August 17 1972, page 334.
25. *Nuclear Engineering International,* September 1974, page 741.
26. *Nuclear Engineering International,* May 1975, page 451.
27. "The Failsafe Risk", by Kurt H. Hokenemser. *Environment 17,* January/February 1975.
28. *New Scientist,* June 26 1975, page 710.
29. "A Poor Buy", op.cit., page 12.
30. *The Case for Solar Energy,* by Peter E. Glaser. Arthur D. Little Inc., Cambridge, Mass., 1972.
31. "The Ecological Crisis of Consumerism", by Alan Roberts, *International,* September 1973: "Crise Ecologique et Soviete de Consommation", *Sous le Drapeau du Socialisme,* nos.60 and 61, 1973.
32. *Nuclear Engineering International,* May 1975, p.447.
33. *Nuclear Engineering International,* April 1975, p.301.
34. "The Nuclear Backlash", by Michael Kenward. *New Scientist,* May 1975.
35. *Nuclear Engineering International,* September 1974, p.743.
36. "World List of Nuclear Power Plants", *Nuclear News,* August 1975, p.63.
37. "Radioactive Waste Management in Selected Foreign Countries", by H.M. Parker. *Nuclear Technology, 24*
December 1974, page 307.
38. "Review of the impact of production and use of energy on the environment and the role of UNEP", by the United Nations Environment Programme, no.75-40793, 1975.
39. See *Nuclear Engineering International:* "Market considerations of medium/small nuclear power reactors", by J. Greason (page 37), and "The case for developing small power reactors", by G. Webb (page 39), both January 1974; "IAEA General Conference asks why no small reactors for developing countries?"
40. "A Critique of Political Ecology", by Hans-Magnus Enzensberger, *New Left Review* No.84, March-April 1974.
41. *"Consumerism" and the Ecological Crisis,* by Alan Roberts, Spokesman Pamphlet No.43, 1974.

The Sizewell Syndrome: Nuclear Power, Nuclear Weapons and Public Policy

Tony Benn

Tony Benn served as Minister of Technology, Minister of Power, and Secretary of State for the Departments of Industry and then Energy, in a succession of Labour Governments. He has thus been responsible for nuclear power for longer than any other minister.

Tony Benn held major responsibilities for Britain's nuclear industry over eight year in a variety of ministerial posts. Here in his evidence to the Sizewell Public Inquiry, under close cross-examination, he explains in detail his considered view. The following text consists of an excerpt from The Sizewell Syndrome, *published in ENDpapers 7/The Spokesman 45, in the Spring of 1984. The full text is available on request.*

Foreword
Sir Kelvin Spencer
(former Chief Scientist
to the Ministry of Power)

Tony Benn's Evidence to the Sizewell Public Inquiry reproduced below criticizes the proposal to build a nuclear power station in Norfolk. But a deeper theme runs through it: the preservation of our hard-won civil liberties in this age of rampant high technology.

Every schoolboy knows of the revolt of the Barons against a tyrannous Monarchy which gave rise to Magna Carta some eight hundred years ago. Since then there have been many such revolts, usually taking the form of a growing groundswell of public opinion which has forced the government of the day to change course.

Once more the need is to build up an *informed* public opinion which will change the policies of our rulers, irrespective of political party.

The threat today comes from the exploitation of the newly discovered source of primary energy, the energy locked up in the atom. Scientists, with war as the stimulus; discovered how to release this. Its first use as we all know was to make two atomic bombs and drop them on Japan. While much of the basic science underlying this discovery was done in England, the development work and the manufacture of the bombs was done by the USA. When the war ended the USA alone among the countries of the world had the skill, the know-how, and the large industrial facilities to make such bombs. Realising the danger to mankind if atomic bombs became part of the armoury of nations, the USA decided to keep their knowledge secret for as long as that could be done. A natural and rational decision I think.

But other countries were not long in catching up, Russia and England leading the race. The English bomb was made in the greatest secrecy, even some Cabinet Ministers at the time knowing nothing of it. The cost was disguised from Parliament by manipulating the annual Votes. This secrecy, most rigorously applied, has bedevilled the situation ever since, and has been carried into the civil application of nuclear energy in a quite unjustifiable way.

With the successful testing of the first English atomic bomb the scientists and technicians who had achieved this challenging task were effectively out of a job. Mass production and progressive 'improvement' of bombs was humdrum work which had no appeal. So they set about persuading those in the seats of power to develop nuclear energy for electricity power stations. By the middle 1950s they gained their objective, and the first nuclear power plant programme was launched. Government sponsorship for this was placed in the Ministry of Power (now the Department of Energy) which, up to then, had had nothing to do with the subject and, departmentally, were ignorant of its scientific background. The generating boards were very reluctant to have anything to do with it, their hands being more than full with bringing into existence an orderly electricity supply industry from the mixed bag of assets handed to them on nationalisation only a few years earlier. This set the then Minister of Power a problem: how to persuade an unwilling industry to embark on nuclear stations? He – and it was he and not his officials – had the brilliant idea of inviting one of the top men who had made the bomb to be the new Chairman of the generating board (which had just changed its functions and its name from Central Electricity Authority to Central Electricity Generating Board). He accepted. From then on a career in the generating boards (English and Scottish) depended on one jumping onto the nuclear bandwaggon.

I happened to be Chief Scientist at the Ministry of Power at the time, and I accepted with enthusiasm the prospect which "ATOMS FOR PEACE" seemed to hold out. Remember Science had emerged from the war with immense prestige, crowned by the success of the atomic bomb. Many scientists, outside as well as inside government service, had succumbed to this adulation by the public in general, I being as overconfident as the rest. We really thought, we in government service and scientists in general, that we had identified in principle all the major hazards involved in this civil use of nuclear power. And we really thought that a large enough programme of research and development together with gradually accumulating experience of operating nuclear stations, would ensure that all these hazards would be adequately coped with. *How wrong we were!*

I left government service at the end of the 1950s in the comfortable conviction that all was going well. The first nuclear station had not yet come into commission. But, as the years went on, snags arose, more and more. Thus I changed from being an advocate for the civil exploitation of nuclear energy to being an opponent.

Tony Benn started his House of Commons career in 1950 and attained Cabinet rank in 1966. In all he has had eight years as Cabinet Minister responsible for nuclear energy. He describes how his views changed just as mine have. In this Evidence he sums up his present views as follows: "I would be happy to see nuclear power phased out and I could not honestly recommend a further ordering programme". I think it is because we have both changed from being advocates to being opposers that he has invited me to write this Foreword. Our reasons for the change may not always coincide. He has had a much wider experience than I have, and has shouldered vastly greater responsibilities. My reasons for change I have set out in a few articles I've written, on request, for various journals. Briefly, it seems to me that some scientists, when they get tied up in some aspect of nuclear technology, lose some measure of intellectual integrity. They do not hesitate to assure the public that sufficient is known about this new science to make sure that any residual risk there may be is small compared with the alleged benefits. (And what standing has a scientist in balancing assumed benefits against risks?). Tony Benn's statement "You have to niggle away and find out if anyone has a vested interest in suppressing information about hazards" applies, alas, to some scientists who, again in Tony Benn's words have "an attachment to nuclear reactors ... as embracing by those who have that view as that of religious conviction."

How far some scientists have slipped in intellectual integrity since the days of Isaac Newton! In his old age he wrote:

"I do not know what I may appear to the world, but to myself I seem to have been only a boy playing on the sea-shore, and diverting myself in now and then finding a smoother pebble or a prettier shell than ordinary, whilst the great ocean of truth lay all undiscovered before me."

This book will, I hope and expect, reach a large public. It will of course evoke disagreement and even anger from those confined to a narrow nuclear energy theology. This is to be welcomed. But let us be charitable to the many intelligent men who have reached high positions in their nuclear priesthood and now, in the last decades of their professional life, find their industry under increasingly well-informed attack. Which of us in such a position would have the courage to recant?

Tony Benn's Evidence for the NUM

The National Union of Mineworkers called Tony Benn to present evidence in the Sizewell Inquiry. The President of the NUM, Arthur Scargill, introduced him.

Your name is Tony Benn?

Yes.

And from 1950 to 1983 you served as a Member of Parliament?

Yes.

In 1966 to 1970, were you Minister of Technology?

Yes, I was.

From 1969 to 1970, did you serve as Minister of Power?

Yes, that is correct.

And in the 1974 Labour Government, did you also serve as Secretary of State for Industry between 1974 and 1975?

Yes.

And from 1975 to 1979, were you Secretary of State for Energy?

Yes, that is correct.

Could I ask you, Mr Benn, to continue giving your evidence by reading from your proof?

Yes. Inspector, in submitting these points for the consideration of the Commission, I shall be drawing upon my own experience as a Cabinet Minister over the eight years when I was responsible, from memory and from notes and papers as accurately as I can, and shall be ready to answer any question that may be put to me, and hope that I might be permitted to expand upon some of these points.

I should make clear that one of the reasons I have submitted some of my evidence in an interrogative form is that, without a research department to assist me in the preparation of my submission, it would seem the best way that I can help the Inquiry is to draw its attention to certain lines of enquiry that would bring relevant facts to light. That is to say I am trying to throw some light on the background of the decision and the inter-relationship between the factors, and these threads are not easy to unravel, and it is not always possible even for Cabinet Ministers to discover the truth. So I turn now to point number 1.

1. The need for Sizewell B

1.1 The basic argument advanced for the Pressurized Water Reactor at Sizewell B, for the introduction of new technology – the PWR – and for the series-ordering of PWR reactors rests upon two arguments that need to be examined very critically indeed before they are accepted.

1.2 These two arguments are first, that energy demand forecasts make it clear that a large programme of nuclear power stations will be necessary, and secondly, that this demand cannot be met economically in any other way, i.e. that PWR is the cheapest system.

1.3 In my submission neither of these two arguments will stand up to close scrutiny, because both are based upon the acceptance of the assumptions that have been fed into the analysis to justify a decision to adopt PWRs that stem from a quite different motivation.

Perhaps I might just at this point expand on one aspect of motivation. Without any doubt the first grounds for the nuclear power programme was the oil embargo in the 1950s. Secondly, in my submission, in 1973 the pressure for a big pressure water reactor programme was because of the power of the National Union of Mineworkers, and more recently the weapons argument to which I shall later refer.

1.4 The simplest way to check the validity of the forecasts that are being submitted to us, is to examine past forecasts that have been made by the Central Electricity Generating Board, over the years, and compare them with actual out-turn. And when I say "forecasts" here, I don't just mean forecasts of demand, but forecasts of performance of nuclear stations which have not lived up to expectations.

1.5 If this is done, it will be seem that the CEGB has proved to have adopted forecasts that were very wide of the mark as, for example, in 1973 they asked for, I have written in my evidence 22, I believe it to be 18 PWRs, that was the figure I should have put there, to be built within the next few years, as compared to 1978 when they expressed great reluctance to order more than one Advanced Gas-cooled Reactor that year, since they then found themselves with excess capacity.

1.6 All these forecasts prove is that the CEGB has always pressed for American style reactors in the long run, i.e. beyond the immediate ordering needs, and has been opposed to building of more British reactors. I think this question as to why that should be the case merits examination.

1.7 It was the same story with the High Temperature Reactor, the Dragon Reactor, which was under development at Winfrith, which the CEGB declined to support, even though High Temperature Reactors do have the advantage of offering the combined heat and power facility, which potentially has great long term value in terms of higher thermal efficiency.

1.8 Similarly, it was the CEGB that forced the cancellation of the Steam Generating Heavy Water Reactor, which had been ordered in 1974, also under development at Winfrith, and which had offered the possibility of being constructed on a modular basis, i.e. without having to up the scale and which had some similarity to the Canadian CANDU reactor, which had proved itself.

1.9 I believe the Inquiry should go into these matters with great care and, if it does, it may well conclude that the motivation of the CEGB is best explained by a desire to drop all British systems in favour of an American system, rather than by the establishment of any real need in terms of domestic energy demand in the United Kingdom.

1.10 A study of past energy forecasts will also reveal that the input from the Treasury in terms of expected economic growth has also proved to be consistently wrong, in that the British economy and the energy demand that would flow from that assumed growth has never justified the forecasts, either in terms of the growth itself or in terms of the energy demands within that growth. This has left the industry with an excess capacity that has actually increased electricity prices to the consumer, who has been called on to pay not only for the electricity that they use but for the unused capacity in the system and, because of that extra cost, this has contributed to reduce the demand still further and push up the levels of excess capacity again. I think this is an important point to stress because the early claims that nuclear electricity would be cheaper have not been justified in the event.

1.11 Indeed, if the Inquiry asks to see the figures over the years it will be seen that the excess capacity, over and above the planning margin, and the very size of the planning margin, exceeds all the electricity produced by nuclear power, and thus makes it questionable, at any rate, whether the nuclear power stations that have been ordered and built over the years have, in terms of contemporary energy demand, ever been necessary at all.

1.12 If the matter is examined still more closely it will be seen that the very high planning margins adopted by the CEGB have in fact in part been dictated by the fact that delays, breakdowns, plant failures and outages – and I cite Dungeness B, Hunterston, corrosion in Magnox and the derating that followed from it – for nuclear power have been greater than for coal powered stations, although there have been some failures of big generating sets, and thus the decision to use nuclear power had, of itself, driven the CEGB to push up the planning margins.

I now turn to…

2. Need, Supply and Usage

2.1 Another argument here relates to the supply of uranium, a substantial quantity of which is acquired from Rio-Tinto Zinc, which derives it from the Rossing Mine in Namibia (a contract actually signed without consent of Ministers), where the political uncertainties are so great that it would be unwise to assume that it will be a reliable source.

I come next to the next point, 2.2 where again there is a misprint. It should be EURATOM, of course, not European Coal and Steel Community.

2.2 Under the EURATOM Treaty all uranium in the EEC has, now, to be acquired under the control of the EEC Energy Commissioner, and Britain has lost the power to purchase its own supplies, or even to control or possibly own them on a strict interpretation of the Treaty. I mention this because many people are not aware that as a result of our membership of

EURATOM we have actually lost this power and require the consent of the Commission for us to enter into contracts to buy uranium.

2.3 I would put it to the Inquiry that one of the dominant factors which explains the policy of the CEGB and the Atomic Energy Authority, which is itself a Defence Agency, in pressing for the PWR is that there are strong military reasons, with which Government is concerned, for doing so, and that to some extent the economic and energy arguments are a cover. I come back to that point later.

2.4 The British Navy uses PWRs in its nuclear submarines, and this has led to a close link with American reactor systems that they hoped would lead to the ordering of PWRs for electricity generation, especially as there has been a virtual standstill of PWR ordering in the United States itself. It is interesting that this is an American system that has not been ordered, as far as I know, in the United States since 1977 or 1978.

2.5 In addition it has only recently become publicly known that for a number of years some of the plutonium which is produced in British power stations has been made available to the United States for weapons purposes, and since the USA has a weapons programme that requires more plutonium than can be produced in the much curtailed US civil power stations they need to get it from Britain, which could continue to supply the plutonium if this country expanded its civil nuclear power programme. I must say that, in expanding this point, that I personally feel betrayed in that I was never told of this arrangement for the trading of the plutonium from our power stations into the American weapons programme.

2.6 To put it more plainly, every British nuclear power station has, or could, become a nuclear factory for the United States, which puts a very different light upon the much publicised argument that civil nuclear power is all about "Atoms for Peace" and is the finest example of beating "Swords into Ploughshares", which are arguments that in good faith I used and can be found in speeches of my own as Secretary of State in the House of Commons.

2.7 In this context it should be pointed out – this is an economic point – that nowhere in the world has nuclear power ever been developed in response to market forces, as compared to government evidence that the coal industry must find its own level in terms of the market. This test has never been applied to nuclear power, nor has any nuclear power been completely paid for by private capital: indeed, it has depended for its origins and its development upon massive State funding, usually motivated by the defence interests of the nation which adopted it, which is true of the United States, of the Soviet Union, of the United Kingdom, of France, of China, of Pakistan. These raise wider questions, but the point I want to make is that it is the one exception, even to the philosophy of this government, which believes in profitability above everything, that nuclear power should be protected from those stern tests.

I now turn to ...

3. New Generating Plant, Capital Costs and Plant Characteristics

3.1 The next series of questions which the Inquiry should examine relates to the true costs of the alternatives open to us.

3.2 I submit that the true costs of nuclear power have never been made explicit, partly because so many of them have been carried in the Defence Budget, which has never been candidly disclosed in respect of nuclear expenditure, and partly because there has never been a proper allocation of these costs to the civil programme. It has always been open to change the economics of civil nuclear power simply by allocating less of the costs for R&D, which are then carried on the Defence Budget.

3.3 The Research and Development of nuclear power was, and remains, immense, and yet the royalties which fell to be paid by the CEGB never took full account of them, for if they had, the cost of generating electricity by nuclear power would, or perhaps I should say could have been shown to have been far greater than by conventional methods. Indeed, the strict comparison would be the historic costs of the mining industry, which are carried by the industry in terms of capital costs, and the R&D costs of nuclear power, which have been wished away in the Defence Budget.

3.4 In addition, the costs associated with the hazards of nuclear power have never been properly allocated, partly because some of these health hazards have been experienced by those who work in the uranium-mines which are privately owned, as with RTZ, in Namibia, where the rigid standards which we maintain under the Health and Safety at Work legislation do not apply. I think it worth noting here that the decision to shift from acquiring uranium from Canada, where we thought it would come from in the late sixties, to Namibia, not only gave the Authority the right to get unsafeguarded uranium, but also, of course, uranium from mines where the wages were much lower.

3.5 It should also be stated that the RTZ contracts are in breach of the UN resolutions on Namibia and thus future supplies under those contracts may not be secure.

3.6 No allowance is made – a new point – either for the payment of compensation to those who live near nuclear processing plants, as at Windscale or Sellafield, were it to be discovered that there is a real hazard arising from the discharge of waste into the atmosphere or into the sea, and given the record of leaks from Windscale and the new anxiety which is very much in the public mind now, the possibility of substantial compensation being required there has never been allocated to the cost of nuclear power.

3.7 Similarly, the decommissioning costs of nuclear plant have had to be estimated because we have had no experience, so far, of an actual example of decommissioning and if the long term costs of storing nuclear waste were to be taken into account, the cost of nuclear power would be seen to be far greater than has ever been admitted. In that context may I come back to Windscale. The very, very serious leak that came to light in March 1979, of the high toxic waste which is the stuff which would be buried in glass blocks for a period of some hundreds of years, that leaked into the ground. I was told at the time it would require the building of a new plant to clear the soil thus contaminated. These are costs which are never included in the figures put forward by the Generating Board.

3.8 Neither, if I am still on costs, are the extra costs of security to prevent the theft of nuclear materials, of which we have already had at least one episode with the theft of 200 tonnes of uranium in about 1968, or to protect the plants or waste sites.

In this connection it must be added that not only is all the necessary security extremely expensive – the atomic energy constabulary; which I established or rather arranged should be armed in early 1976 – but it carries with it a direct limitation on civil liberties for those who work in or around all nuclear installations, and to that extent adds to the cost in terms of the quality of life of all concerned.

3.9 None of these disadvantages apply to coal, oil or gas-fired electricity generating stations, nor would there be costs on a comparable scale. But even without those factors, the estimates made for me established that if like is compared with like, truly like with like, there was no real difference in cost between a coal-fired station and a nuclear-powered station, so long as both were used as base-load stations on a comparable place in the merit order, and this related to AGR stations which are likely to be less expensive than PWRs. The reason I say that is that Pressurised Water Reactors, redesigned for British conditions, would require modification to meet higher British safety standards and the costs for those modifications have never been established. I understand, Inspector, that the final safety clearance for the PWR has not yet been granted by the Nuclear Inspectorate. Correct me if I am wrong.

3.10 The safety factor relates to the different design features of the PWR, the doubts about the pressure vessel and certainly when Sir Alan

Cottrell came to see me as Chief Scientist with Dr Marshall, Sir Alan Cottrell expressed himself then of the view that the original Pressurised Water Reactor safety studies which Dr Marshall had conducted did not satisfy him. Therefore, in a sense, even if the safety modifications are made, this will not alter the basic case that the design features of the PWR render it unsafe.

3.11 But quite apart from the comparisons with base-load coal stations, the most economic system now – I am speaking short-term – would undoubtedly include the refurbishing of existing coal stations, where you have existing transmission lines, and the development of combined heat and power using fluidised-bed burning in the longer term, which offers greater fuel efficiency and lower pollution. Clearly, any big station, even a coal fired station, which discharges a lot of its heat into the atmosphere, has a lower thermal efficiency than if you can build a power station or refurbish a station within a town where you can use the heat as well.

3.12 In terms of the most economic return on investment it is also clear, it is self-evident, that conservation brings a far quicker return than a new nuclear power station and, I might add, creates more jobs, because there is no wait during construction when you are doing conservation and virtually no recurrent operating cost. If you insulate somebody's home on Monday, the savings begin on Tuesday. If you are spending ten years building a power station all you get at the end is electricity.

3.13 Comparable capital investment in renewable sources and other benign energy sources will also bring a more rapid return and will not suffer from the health and safety hazards associated with nuclear power.

In this context may I mention the Severn Barrage in particular, because I would invite the Inquiry to look very critically at those who say the Severn Barrage is not economic on present showing, because at the very moment when the present Severn Bridge is under hazard from the point of view of safety, the Severn Barrage would give you a second river crossing. This would be to reduce the cost of the Severn Barrage for generation purposes. I believe the Severn Barrage has acquired a relevance as an alternative energy source much greater than would have been the case even a few months ago.

3.14 It had always been argued that these alternative resources would not be available until the turn of the century, but with the passage of the years this date is getting closer and the delays in building nuclear plant means that a major new nuclear power programme could not be in operation until a comparable date.

Again, if I refer to my experience at the Department of Energy, ten years ago Officials of the Department of Energy, and I can document this if required, were actually speaking about the first PWR being operational by 1980/81. Here we are in 1983 with no prospect of even completing the inquiry quickly and then the period of construction. So the chronology of timescales between alternative energy and nuclear power shift strongly in favour of the former.

3.15 What matters now is that the public investment should be shifted towards coal and conservation and alternative energy sources to cut the delays that would occur if these investments were held back so that nuclear power could go ahead.

In this context I would draw your attention, Inspector, to three points. One is the enormous subsidies to coal in the Common Market compared to the very meagre or non-existent ones in Britain. Secondly, the existence of the enormous reserves of coal in the North Sea which, with technologies not difficult to contemplate and develop, you would be able to scour these coal seams and pipe the coal ashore as slurry. Therefore, it is probably true to say we have one thousand years of coal and not three hundred years, and this coal for oil-conversion. For gas, for a feedstock is of great value.

Thirdly, as you may have noticed recently, a Government report said that after a nuclear war Britain would depend upon coal and therefore special shelters should be built for mining engineers so they survive a nuclear war, which indicates that even the Government, wearing one hat, thinks coal has a future.

4. Thermal Reactor Systems Available

4.1 Britain has pioneered and developed and operated Gas Cooled Reactors from the very beginning and has no experience of the operation of PWRs for electricity generation. I leave aside the submarines.

4.2 The Magnox stations have provided power, although corrosion led to the de-rating of some stations, particularly after the corrosion came to light in 1969/1970.

4.3 The AGRs came on stream later – Dungeness B, the first ordered and virtually the last on stream – but the AGRs have extended our experience of gas cooled systems and I have heard it argued seriously that it would be unwise for the world to become entirely dependent on the PWR, given its uncertain safety record at Three Mile Island and elsewhere.

4.4 Britain also has the fast breeder station at Dounreay and is working on JET, the fusion project at Culham in association with other countries, notably the EEC.

4.5 It makes no sense for a country of our size, which is in any case –

alone among the western countries – self-sufficient in oil and gas and has 300 years of recoverable coal reserved to add a fourth nuclear system to its present range.

I would like to emphasise this point. We have suffered substantially as every one of the Magnox and the AGRs – or almost every one – was a one off. We did not take advantage of the benefits of replication. You would then have to come down the learning curve again with the PWR, at a time when the case for it is not established.

4.6 The world estimates of the installed capacity of nuclear power stations, made 10 years ago, have been sharply cut back, with the United States experiencing a virtual freeze – I believe something like 45% of their orders have been cancelled – with the EEC cutting its early estimates by about half, with the cancellation of the big Iranian programme ordered by the Shah, which was when Mr Marshall visited the Shah, and the Shah offered to buy half our nuclear industry if only we would order a PWR, and the sharply reduced demand for energy in Britain.

4.7 There is no doubt that the failure of the United States nuclear power companies to win orders in America lies at the root of the fierce lobbying for orders in Britain, together with American weapons needs, creating the single most powerful industrial and political lobby in the UK that I have ever come across in my Ministerial experience, including in particular pressure for us to move to the Westinghouse system with whom negotiations had actually been going on in the early 1970s.

4.8 For Britain to launch into a major PWR programme at this stage cannot be justified on any rational basis, and the Department might consider seeking more information from the CEGB on this question, especially the old CEGB arguments, and the National Nuclear Corporation arguments in particular, that Britain would be able to enter the world market with a PWR. This is an argument that simply cannot be sustained with the Americans, German and French already in the PWR business and virtually unable to get orders. But this argument was certainly brought out in the Department of Energy brief given to the incoming Labour Minister in February-March 1974, and was one of the major arguments put to me when I was asked by my officials to adopt the PWR and refused to do so in November 1977.

So I come to ...

5. CEGB Thermal Reactor Strategy

5.1 This strategy has been, ever since 1966, to build up the nuclear power programme wherever and whenever the Board thought that it could persuade the Government of the day to adopt an American reactor, and to hold back on the ordering of AGRs or coal-fired stations if these were pressed. I mean, clearly, you had to have some other orders since the PWR was not available, but the view of the CEGB depended very much on what system they were offered. If they were offered the one they wanted they would want a lot. If they were not they did not.

5.2 It is a strange commentary upon the official attitude to our own science, technology and operating experience that whereas the AEA should always have urged Government to invest heavily in nuclear R and D to design and develop gas cooled systems, and indeed, the fast breeder, the CEGB should have for nearly 20 years, consistently urged us to abandon the systems that we have developed, and that applies to the HTR, to the Steam Generating Heavy Water Reactor which was both ordered and cancelled, to the AGR and indeed even on the fast breeder, to go in with the French on the Super Phoenix which was a view which was held in 1975.

6. Conclusion

6.1 In conclusion, Inspector, I submit to the Inquiry that the case for the PWR cannot be sustained on energy or economy grounds, that the PWR is the wrong system for Britain, even if a limited nuclear power prgramme is to be sustained.

6.2 I also recommend that the right course for the Inquiry to recommend itself would be to step up the investment in coal, conservation and alternative sources of energy, and to urge that the necessary investments be put into them. Thank you very much indeed.

MR SCARGILL: *Inspector, you will note that in the foreword to Mr Benn's proof of evidence he expressed the hope that he would be permitted to expand on the points that he made. Arising from the very significant arguments that he has advanced there are a number of questions which I would like to put to Mr Benn, some of which deal with very sensitive areas and which could possibly impinge on both the 30 year rule and the Official Secrets Act. I am therefore asking you for your agreement that I be allowed to put certain questions to Mr Benn in order that this Inquiry shall have the benefit of information from a person who was Secretary of State for Energy at a very crucial stage of the whole debate on nuclear technology, and in particular when the issue of the PWR was being discussed.*

THE INSPECTOR: *Mr Scargill, you were kind enough to give me notice that you were going to raise these two points. I am anxious that wherever possible all evidence that is valuable to the Inquiry shall be given. There are two difficulties which I would like you and Mr Benn to bear in mind, however. The first is, one of the reasons for asking for proofs from everybody and to require them to be deposited in good time is so that all parties would have an opportunity to consider what questions arise from them and the opportunity to put them fully and clearly and well, particularly where they involve technical and documentary matters which need to be checked. It follows, therefore, that if Mr Benn raises points that go beyond what other parties could reasonably have anticipated we may have to delay questions on them.*

As to sensitive issues I am not, of course, a free agent on these matters. I am bound by the rules which govern the Inquiry, and so are we all. I am concerned, particularly in the light of the way Ministers put their descriptions of the importance of the Inquiry, that wherever reasonably possible we look at facts or material which may help, but we shall have to watch very carefully to see whether it passes the line that I have to observe, as the rules are quite clear that I shall not permit certain matters to be put in evidence, but we shall have to see how we go. Certainly so far we seem to have kept well within the rules. Let us see whether we can continue to do so.

Subject to that, yes, please, put the questions you have, and we will look at each of them as they come along.

MR SCARGILL: *Mr Benn, could you first of all explain to the Inquiry the primary functions of the Department of Energy, and if possible its relationship with other Government Departments, particularly when talking about new energy development?*

Yes. Well, I will do my best. The Department of Energy inherited the statutory duties of the Ministry of Power, which was a position I held earlier, and of course, new power sources have come on, notably nuclear power itself. It has responsibility for the co-ordination of energy policy, and during the period of my occupancy of that office we set up the Energy Commission, and we published everything. We published the transaction of the meetings of the Energy Commission. We published all the documentation. Therefore, during that period at any rate it was very open and clear what was going on insofar as it related to energy policy as such.

The Department relates to other Departments in obvious ways, and in some respects less obvious ways. The Treasury, of course, is responsible for clearing the budget of the Department. The Foreign Office is responsible for its international relationships, i.e. with the International Atomic Energy Authority, with the EEC, with the American Government where nuclear relationships lie at the very core of the special relationship. The Ministry of Defence is passionately committed to nuclear power for the reason I gave because I think their main reason is they need the plutonium themselves. The Department of Employment has responsibility for the Health & Safety Commission to which responsibility for the Nuclear Inspectorate was transferred, in my view quite rightly, to get it out of the hands of the Department of Energy. The Department of the Environment has responsibility for waste management, which was a move I recommended again to get it out of the hands of the Department of Energy, and of course the Scottish Office is involved and so on.

So the Department of Energy lies at the heart of a whole network of other Departmental responsibilities, and to that extent its major decisions have to be collectively agreed by Cabinet and cannot be decided by the Minister alone, quite properly.

MR SCARGILL: *In view of your previous statement would you say that as a former Secretary of State for Energy you were at all times advised correctly and accurately by those within the Department on matters relating to either the pressure water reactor or any other section of the nuclear power programme?*

Well, if you come to the pressure water reactor, the position was that the Cabinet decided and I announced in January 1978, that an option should be opened up for the pressure water reactor. It was a difficult argument to resist in theory. The Generating Board at the time wanted this to mean a commitment to order, which was a view that I personally did not share. As far as the PWR, therefore, is concerned we had no direct information, but it so happened at the very end of my period of office the Harrisburg accident occurred and I did write a manuscript amendment to a letter to Sir John Hill telling him to stop work on the PWR until the Inquiry on Harrisburg was available, and that letter was actually sent to him and then withdrawn without my knowledge by the officials. But that was the view I took.

If you go back on to other questions, I must divide them into two matters; one in respect of military information. As I made clear in my evidence, I was not aware that we were supplying plutonium for American weapons purposes, and I think I should underline the point that on matters relating to Defence the public should not rely upon the truth being told by Government, nor should Ministers rely upon their being fully informed.

On other matters, it was sometimes difficult to get information. For example, some of the incidents that occurred were not brought to my attention. The big Soviet tragedy in 1957 at their reprocessing plant was not brought to my attention, and when I discovered I learned later that the AEA had known about it but had been asked not to tell Ministers. Secondly, there was the theft of 200 tonnes of uranium in about 1968 which was not brought to my attention, and when I asked why they said because it was a matter for EURATOM and we were not in the Common Market. I had difficulty in discovering about some of the leaks at Windscale. I must say, candidly, it is very difficult even for Ministers to get all the information necessary, and this does make it hard, therefore, to retain and develop democratic control of nuclear power. This is a point made very often in public. I am not making it for the first time now.

THE INSPECTOR: *You are so far closely following the evidence you gave to the Select Committee, which we have read.*

MR SCARGILL: *There are, however, two points arising from that answer, Mr Benn, that I am sure must give concern to the Inquiry. The first is that you say a letter you sent to Sir John Hill was withdrawn. I would like to know and I am sure the Inquiry would, by whom, and secondly, the fact that you state as Secretary of State you were not told of the disaster in the USSR. I would like to know who gave the instruction that you should not be informed?*

THE INSPECTOR: *Well, Mr Scargill, I think this gets beyond what is really helpful to me. I do not know what Mr Benn thinks about it. The points he has just made were drawn to my attention before the Inquiry began. We have all had them very clearly in mind. Mr Benn is right in saying he set them out very lucidly and clearly, and those are factors of course I will bear in mind. As to who gave an instruction does not seem to me in any way to help whether I should make the recommendations Mr Benn suggests I should make or whether it does not. So I would want some persuasion that we should go into an area of that kind on any basis.*

Could I assist you, Inspector, on this point? The first point about the Soviet disaster and so on is on the record in my evidence to the Select Committee, which I hope perhaps you will accept as a document in the Inquiry and does cover these points. It has never been brought out because I have never been asked before what actually happened in April 1979 in respect of the letter I sent to Sir John Hill, but I do believe it is significant if I may put it to you (and I do not have to identify the name of anyone involved) that I did think it necessary after Harrisburg to write to Sir John Hill, and I did it in the form of a manuscript amendment to a letter, telling him to suspend work on the PWR, and I wrote to the Prime Minister and said I had done this. I think it is relevant that an official – it does not matter who – having despatched the letter sent somebody to collect it so that in a sense that letter was never on the records of the Atomic Energy Authority. I think that does highlight one of the problems, namely the problems Ministers have where in pursuit of what they think to be right they may run across the policy of the Department. It so happened this all occurred within a couple of weeks of an Election and therefore it was overtaken by events, but it is an important bit of evidence, I think.

MR SCARGILL: *Inspector, could I make two points by way of a submission on this: it is, I contend, of vital importance that we do know these details, even if that means producing documentation which must be available in ministerial offices, because we are being asked to accept a new technology, and it would appear from the answers given to questions that I have just put to Mr Benn, that the Secretary of State responsible for that Department is not advised properly, does not have proper knowledge of events surrounding the industry or industries over which he has control, and furthermore, it would appear, most astonishingly of all, that someone, or somebody, is actually in a position to give orders to stop information going to the Secretary of State.*

Now all that does relate to cost and civil liberties, and I suggest that these matters are of vital importance to this Inquiry.

Could I suggest that if these questions arise – and they are quite proper questions in my opinion – the right thing to do would be ask the Deparment to disclose the documentation, not for me to do so, because that letter to Sir John Hill will be on the record, and also the response to my enquiry as to why it had been stopped and returned, but if I could divert questions relating to the disclosure of documents to the Department rather than try to take responsibility in any way myself, I think ...

THE INSPECTOR: *Well I share your view, Mr Benn. I think that is* right. Now can we go on to the next question please, Mr Scargill.

MR SCARGILL: *I note the point that at least it will be considered pertinent that documents should be produced.*

THE INSPECTOR: *Well I accept the point as put forward by Mr Benn, and in no other form at the moment.*

MR SCARGILL: *Mr Benn, during your period as Secretary of State, did you have any close relationships in terms of demand for electricity or other sources of energy, with the Chairmen of the Energy Industries, as a collective body, and as individuals, and could you tell us what your experience was?*

Well, I have touched, I think, on this point, but I would like to draw it to the attention of the Inquiry. The powers of a Secretary of State in relation to the Central Electricity Generating Board are very inadequate and remote. My own draft bill, which never went through the House because it was too late in the Parliament, dealt with the matter, but I must tell the Inquiry that a Secretary of State does not even have legal power over the Generating Board, except in respect of what are called 'general directives'. Now general directives are so general that nobody has ever been able to frame one general enough to be legal, and actually what you wanted was information.

Now if I give one example which is of particular relevance to you. In 1972 or '73 the Generating Board ordered coal from Australia on the grounds that it was cheaper. I think it was primarily ordered to undermine the bargaining power of the NUM, but that is irrelevant. By the time the coal arrived, the pound/sterling was so much lower in value, that actually it was more expensive from Australia than it was from Britain, so what the Generating Board did was to sell the Australian coal at a loss to the French Electricity Authorities, thus imposing a burden upon the industry, not due to the cost of British coal, but to the failure to order properly, indeed to order at all, and when I asked Sir Arthur Hawkins what was the price he paid for the coal, he said to me quite bluntly and plainly, because he was a blunt and plain man: "That is a management decision. You have no right to ask me", and he actually laid upon the table between us the statute which he claimed authorised him not to answer the question. So there are problems of a practical character in management matters, vis-a-vis Secretaries of State in nationalised industries, and I think it is important that people should know, because I believe the legislation should be altered so that ministers are actually in charge of these great corporations when matters as important as this come up.

MR SCARGILL: *You have touched, during your evidence, on the question of defence and the sale of plutonium. Could I ask you if you were aware that Britain was selling plutonium and to whom, and what effect this sale had in terms of its economic impact on the industry?*

Since I did not know about it I cannot comment on the economic aspects, but I had no knowledge that we were selling plutonium to the United States for military purposes. Now that this information has been brought to light by Dr Hesketh, and I think has been publicly admitted, I do not think there is any argument about it. I have made enquiries myself and I believe it has been public knowledge in the United States for some time, so though it was a disclosure here, it was not so much a disclosure there, but of course one can then see wholly different motives for many of the decisions that at a time were presented as being purely economic in character. I believe, for example, PWRs produce more plutonium, and that would meet the hungry American weapons programme which cannot be met by their own civil nuclear power stations, and I think that must have been a factor. And indeed when I analyse again - because I have gone over every one of these experiences time and again to try and understand what really happened - there are so many mysteries about why a PWR that really had so many obvious disadvantages should be pressed with such passion, if it was not that there were other factors. I believe the defence argument was a factor that must have influenced the Cabinet office who would have known, and the Ministry of Defence, not known to me, and the economics of it would all be adjusted to suit whatever the Ministry of Defence wanted. I mean there is no concept of economics when you are dealing with military matters. If they want it they spend it, and if they want to cover it up by taking the cost on their budget and making it look as if the PWR is cheaper, that is what they do, and I think one should not be under any illusion that these decisions are taken by meticulous examination of computer printouts. We live in a real world and I think reading some of the economic figures that have been put in, without disrespect to economists, they are living in a world that does not actually relate to the way decisions are taken. Decisions are taken on a practical basis according to the objectives you have, the options that are practically open, and whether you can afford it or not, so I think these military aspects probably played a much larger part than has ever been brought out in any of this sophisticated calculation about return on capital that pour out of the CEGB with the regularity of a programmed computer. The thing does not quite work like that.

MR SCARGILL: *What you are really saying, Mr Benn, is that the statistics and the figures consistently put forward by the CEGB, and indeed by the Government, bear close scrutiny because they often conceal rather than reveal?*

l am not suggesting there is concealment in statistics. What I am saying is that statistics reveal the assumptions that are being fed into them. You can feed any assumptions in and get any results out, and what I am saying is – and I think it is important since this is a practical Inquiry – we are not in a Court of Law or qualifying for a mathematical degree at Cambridge, that people should actually understand that when a Government sits down to decide what to do, if it wants the bomb, then it will take into account the need for the bomb in its decisions about civil nuclear power.

If, for example, the power plant industry desperately needs a new order and if it does not get it it will go out of existence, Government will take account of that. If there is a danger that a capacity to produce nuclear power would disappear without an order, Government will take account of that.

The only point I am making, because I am not suggesting that anybody is misled in the statistical field, is that the thing has got, in my opinion, much too sophisticated in the presentation down to three points of decimals, whereas actually ministers are serious people and so are civil servants and generals and Generating Board people, and they sit down and say: "This is what we would really like. Let us see how we can present a good argument that makes it looks as if that is the only course we can pursue." Now that is real life, that is all I am saying.

MR SCARGILL: *In your proof of evidence, you stated that the decision in 1978 to explore a PWR option was a Cabinet decision. Could I ask you what recommendations or advice was given to you and your Department, and by whom?*

I do not want to make too much of a meal of it, but it was the first time in my ministerial life that I had a meeting with all my officials, and they were unanimous under the Permanent Secretary we should adopt the PWR. I declined to accept that, and as a result they declined to draft a paper for me to say what I wanted – this is in the public record. It was a very extraordinary example, and so I got my advisers with the help of one official who was prepared to put his career at risk, and my Department then made its paper available to the CPRS, Central Policy Review Staff, who put it in as a paper of their own. But in the end when it came to Cabinet, the weight of argument was so powerful that we did actually carry the day, i.e. the Cabinet did not accept the PWR option, and it was left with this slightly fudged compromise that we would keep the option of a PWR open, which was a difficult thing to resist, but fall short of a definite order. Although the CEGB at the time wanted a definite order, in order, as they said, to give credibility to the examination of the option; so that was the way it came out of the Cabinet, but it was the most interesting example of a Department in effect going on strike against its Secretary of State. They simply would not prepare a paper because they did not agree with the decision that I had reached.

MR SCARGILL: *Did you at any time form an opinion about any relationship that was going on at the time, or may even be going on now, between the Department of Energy and any one of the corporations who have designed or built pressurised water reactors?*

I think there is something a bit funny, I will be candid with you, and I have got to be careful. I think there is something a bit unhealthy about the relationship between Westinghouse and some of the people operating in the general area in this country.

For example, in 1975 at the time of the Referendum when ministers were busy, some sort of an arrangement was reached with Westinghouse, which I was told meant we were committed to them. When I examined the papers very carefully I found it was not quite like that, and actually I think Kraftwerke Union in Germany, whom I saw about it, would have probably offered a better bet if you were to go along that road.

I think there is an unhealthy relationship, and I refer to that in a very general and discreet way in saying that this is a very powerful lobby, extremely powerful lobby, and it is powerful in part because it has such formidable allies within Whitehall. It is not an external lobby, it is an external lobby with very close internal connections, for all sorts of reasons I can understand, but it does mean a minister trying to arrive at a decision, as I was, that was in the best interests I thought of the energy industries and energy policy, found it very, very difficult to contend with.

MR SCARGILL: *What you are really saying is that there are relationships* between people in the Department of Energy and an organisation like Westinghouse, which are about to have, or create an imbalance in terms of input to any Secretary of State for Energy?

Well I do not want to throw doubt upon the integrity of any individual, and I do not think that is perhaps the right way to say it. What I am saying is that there are some very strange relationships in this whole area which may only be explicable in terms of things that have never come to light as, for example, the military side. I mean if I had known in 1977/78 what I now know about the sale of plutonium for American weapons purposes, I would have conducted the whole argument completely differently, because I would have recognised that there was a military element which I was fully unaware of, and that I would have examined some of these arguments in terms of the military aspects, what plutonium is generated by an AGR as compared to a PWR and so on. I think it is that, it is the need for greater illumination of what is happening, more light rather than any suggestion of people behaving improperly, and I would not want to give that indication.

MR SCARGILL: *Let me put it this way ...*

THE INSPECTOR: *Mr Scargill, I think I have this point now, and I would like you to move on to the next one.*

MR SCARGILL: *Could I ask you then, Mr Benn, were you ever told of a theft of uranium in 1968?*

No, I was not and I was very angry when I discovered that there had been 200 tonnes of uranium stolen, and I asked my officials why I had not been told because I was the Minister. In the Soviet case, of course, I was not the Minister and I was not told because apparently the CIA did not want Ministers to be told when they discovered it. That is what I heard. But in the case of 1968 theft of uranium, I was not told, and when I asked why not, I was told that we were not a member of EURATOM at the time and therefore they did not want to worry the minister. It was rather like the television programme 'Yes Minister. No Minister', only it was dealing with some pretty big issues, and I did feel incensed on that occasion that I had not been informed.

MR SCARGILL: *I have got three more questions. The first one is: are you aware at any time of any discussion, or pressure, or dialogue, between a*

British Prime Minister, a foreign Prime Minister, which would lead them to believe that irrespective of a Cabinet decision, Britain would go ahead with a pressurised water reactor programme?

THE INSPECTOR: *Well, Mr Scargill, this is getting beyond what direct evidence can provide, and I do not think that is really going to help me.*

I think disclosure of documents, and I would base myself, Inspector, if you would agree, that it would be a very good thing. Frankly, if people knew more about what really went on, and I think if I leave it at that, i.e. discussions about relations with the United States in nuclear matters and relations with the Common Market and so on, because I think the disclosure of documents here would throw a different light on some of the decisions that have been held.

MR SCARGILL: *May I say, sir, that you have been extremely fair, and could I just add for the record that I would certainly like to have the disclosure of these records, because information has come into my possession which is extremely disturbing, to say the least.*

THE INSPECTOR: *Well I note your point.*

MR SCARGILL: *Could you tell me, Mr.Benn, what your attitude is, or view, of a Public Inquiry, and what its purpose is.*

Oh I think the Public Inquiry of the kind of which I am now attending and giving evidence, is of enormous importance. We had one on Windscale, which was the first of these proportions, but I believe that it does a whole range of things. Its prime purpose, I suppose, is to satisfy the people in the area that there is justification for a project that would be sited there, but in the event, of course, it has become an opportunity, quite properly, to examine the thinking and policy and alternatives open to the Government when such a proposal is made. It provides, if the Inspector is tolerant, an opportunity, therefore, for evidence to be brought to bear on a wide range of relevant matters. It educates the public. It puts Ministers and Officials under pressure, and I think it alters the course of events. I hazard a view that the pressurized water reactor will not be built in Britain. That is my conviction, and I believe that whatever the recommendation may be, and I am not anticipating it, the evidence brought out in this Inquiry will greatly confirm the view of those who think that it should not be built. I think they are very important indeed, and I do not believe there is any other country in the world that has this type of procedure, and I think it is very much to our credit that we have it.

MR. SCARGILL: *Could I ask you to comment, if I may, on a piece in yesterday's* Guardian, *dealing with the Public Inquiry into the Sizewell B reactor. It refers to Sir Walter Marshall, the Chairman of the Central Electricity Generating Board, speaking about this Inquiry. It is quoted in the American magazine* Forbes, *and he says:*

"I expect to get approval in about a year's time. By that time the British public will be bored to tears by nuclear power. That, of course, is the purpose of having a public inquiry."

THE INSPECTOR: *Well, Mr Scargill, that is a matter of someone else's view. I think I have Mr Benn's view about the matter very clearly in mind.*

That is a vintage quotation, if I may say so.

MR SCARGILL: *One final question, Mr Benn. In what sense have you changed your views or your attitudes in respect of nuclear power over the past few years?*

Well, it is a fair question. I had the responsibilities over a long period and obviously, in preparation for this, I read most if not all of the speeches that I have made. I think anyone who did the same, and I am not urging you to do so, would find that I have always laid great emphasis on candour and openness and the disclosure of information, from the very beginning.

I do not believe decisions of this magnitude, involving risks that are quite unparalleled in the history of energy policy, outside military weapons – there is nobody else to take such important decisions as we are about nuclear power for civil purposes, and I believe that the maintenance of a democratic control of this technology is very important. But it is true, having said all that, that events have contributed to convert me from, in the first place, in 1966, being an honest and open advocate of Atoms for Peace to the position where I now feel nuclear power should not feature in long term energy policy. I have come to that view slowly – I think in many ways too slowly – and if you are going to have a programme at all, it should be a minimal programme, which is what I announced in 1978. But, I do not personally believe, in the light of what we now know, that this is the right road to follow. I am strengthened in that by the fact that the Americans, who are well ahead of us, stopped it so many years ago.

I think what influenced me, if you ask me the question, was the Windscale leaks, particularly the latest one, which in 1979 released the high toxic liquor, as they call it, offering to people 500 to 1000 rads an hour, whereas the legal limit is 5 per year – the really staggering danger of that leak; I think what happened at Harrisburg and the incident over my letter, to which you made reference; I think the decision to go ahead with the PWR, which is quite the wrong decision; I think the discovery, if I can broaden it without developing it, that the Chevaline military weapon was ordered without telling the Cabinet; I think experience over the Pakistan bomb, where we discovered they were building the bomb and we stopped supplying to them, and then when Afghanistan was invaded Pakistan was renewed, so, the non-proliferation controls do not exist; and I think, finally, the discovery that having had those responsibilities for eight years, I did not even know about the plutonium deal.

I think these are the factors, plus civil liberties and so on, which have inclined me to the view that we have really been misled about the potential safety and value of nuclear power, and I agree with Mr. O'Leary, the former President of the Federal Energy Commission and Assistant Secretary to Schlesinger, that in 100 years, probably, there will not be civil nuclear power. I think those are the factors that have shaped my thinking, if I can summarise them very briefly.

MR SCARGILL: *Thank you very much indeed Mr Benn.*

"Civil Nuclear Energy Is Neither Safe Nor Essential"

Petra Kelly

Petra Kelly was spokeswoman of the 'Greens' and a member of the German Parliament in the 1980s. This is the text of her speech at the Oxford Union in June 1986.

This text was published in ENDpapers 13/The Spokesman 51 *in the Autumn of 1986.*

Since the time of my sister Grace's death from cancer many years ago, I have worked with children suffering with cancer, with doctors treating cancer patients, and with the families of the patients. In this way, the implications – medical, political, social, and economic – of civilian, military, and medical uses of nuclear energy have been for a long time a part of my political and personal interests.

I am greatly frustrated by the mindless assurances which follow every nuclear accident or radiation spillage that there is "no immediate danger". After Three Mile Island, and still more so after Chernobyl, I hope we are better able to grasp the implications for human health of all radiation exposure, and the reasons why we must reject nuclear technology. I truly believe that one day there will be a global consensus against all civilian and military nuclear options. There are also risks with some of the other energy systems, but nuclear energy risks irreversible damage, both political and genetic! All governments, together with the nuclear lobbies in East and West and in the northern and southern hemisphere, seem to share the same destructive behaviour; it is this behaviour which must serve to mobilize us in non-violent protest against all uses of nuclear power.

In the developing world there are many constraints to life and growth such as hunger, illness, poverty, and repression, and they are all clearly interrelated with many of those in the developed world such as unemployment, cancer, leukemia, and the nuclear threat. We must create soft ecological technologies and non-violent energy strategies in harmony with life and earth, and a new and just social order, which assures the world's children a just and ecologically balanced future.

Before stating our case against nuclear energy, I should like to begin on a more positive note by putting the case for non-violent energy planning. We need a soft path to peace and a soft path to energy planning.

We must eliminate the attraction of and vulnerability to attack and sabotage of energy sources. We must reduce the international inequities that lead to wars and military interventions. We must eliminate those energy technologies that provide the means to make nuclear bombs and provide "innocent" disguise for "bomb factories". Most nuclear plants can be converted to bomb factories and thus serve as ambiguous threats that lead rivals to wish to possess their own bombs. Non-violent energy strategies are not only important for us in the developed countries but for the developing ones as well. They should emphasize diversity of energy sources in order to provide local control and responsibility; dependability of energy supply to prevent catastrophic losses of power; development of energy technologies based on local resources which make best use of human skills and prioritise "least-cost" energy sources to conserve scarce capital and to reduce opportunities for corruption. Non-violent energy strategies mean the elimination of nuclear power.

Until now we have been rendered psychologically "numb" by the civilian uses in order to accept the bomb and nuclear power. I believe that the psychological climate is growing for an end to international political and nuclear blackmail and violence, and for an end to the criminal deeds of governments. As Theodor Roethke once remarked: "In a dark time, the eye begins to see."

The nuclear genie cannot be put back into the bottle. And even if it could, it would not change the fundamental problems that set human beings one against another: power without purpose, paternalism, chauvinism, sexism, racism, and injustice. But a soft energy path would foster a social framework in which to address these problems, and would eliminate one of the weapons that exacerbates them. Even if the nuclear genie is finally controlled, reduced to a large extent, and eventually abolished, there are other things still waiting in the wings: nerve gases, space weapons, SDI, lasers, germ warfare. Atomic weapons are only one part of the arsenal amongst several. But after Chernobyl we have all the more hope of abolishing the nuclear industries. It is important to know when to say "No!", to recognise the limits to an industry where no human errors are allowed, and where irreversible damage can occur. I would here like to cite the words of General Omar Bradley, himself hardly a pacifist. In a book called *The First Nuclear World War* (by Amery B. and Hunter Lovins), he stated: "Ours is a world of nuclear giants and ethical infants. If we continue to develop our technology without wisdom or prudence, our servant may prove to be our executioner." Civilian and military nuclear energy are already our executioners.

Reinterpreting Article IV of the Non-Proliferation Treaty and initiating non-violent energy strategies in developing and developed countries remain the only way to remove the ambiguity in this treaty, which has allowed and indeed encouraged the proliferation of nuclear weapons. Any country that embarks on a non-violent energy strategy and gets rid of nuclear power will show its neighbours and friends, as well as its opponents, that it is not involved in clandestine bomb-building. Let us not forget that nations with nuclear facilities will run the risk of pre-emptive strikes by their neighbours and enemies who suspect them of "nuclearizing". Within NATO strategy we know that conventional bombing of nuclear power plants will cause catastrophic fall-out, and this is one argument why we must reject nuclear power plants in such densely populated areas as Europe, for example. We don't even need nuclear weapons to make a nuclear holocaust; all we need are nuclear power plants and conventional bombs. And such a war will be hell for us and for future generations. In fact, NATO once stated that Europe is *not defensible* due to the number of nuclear reactors that could be targets.

For 30 years now the nuclear power industry, especially in Western countries, has sold to third world countries the material and the know-how to build nuclear weapons. By 1990 the list of countries able to produce nuclear weapons will include Israel, Iraq, Libya, India, Pakistan, Taiwan, South Korea, Brazil, Argentina, and South Africa. I believe that at least half of these can already do so today. Civilian nuclear power provides the means and the "innocent cover" for making nuclear bombs.

Nuclear power cannot survive in a so-called "free market" and I believe its impending global economic collapse offers a unique if brief opportunity to cut energy costs and to stop nuclear proliferation.

The peaceful and military uses of nuclear energy and technology are fully intertwined; they are Siamese twins. The ashes and agony of Hiroshima and Nagasaki provided painfully vivid proof of the destructive potential of atomic energy, while the so-called "peaceful" or civilian uses of this new source of power were nothing but forecasts and speculations at the beginning. In 1968, the priorities had already become completely reversed and US Ambassador Arthur Goldberg told the United Nations in a speech in support of the Non-Proliferation Treaty that it would be an "unacceptable choice", indeed "unthinkable" to decide that the non-nuclear countries "must be without the benefits of this extremely promising energy source, nuclear power – simply because we lack an agreed means to safeguard that power for peace." During the 60s and 70s, the so-called civilian "nuclear ploughshares" were to be distributed throughout the world, even if they could easily be forged into swords.

The story of the International Atomic Energy Agency and of EURATOM is a sad one. The philosophy behind the controls tilts towards the potential beneficiaries of nuclear assistance by stipulating that safeguards must *not hamper the flow of nuclear know-how and materials claimed to serve "peaceful" ends.*

Thus, for example, Article 2 of the statute of the International Atomic Energy Agency states that the Agency, "shall seek to accelerate and enlarge the contribution of atomic energy to peace" and "shall assure so far as it is able that its existence is not used to further any military purpose. But if the agency *is not able* to ensure this, how can it continue to accelerate the spread of nuclear technology?

Similarly, the Non-Proliferation Treaty (Article III,3) says that the safeguards must "avoid hampering" international co-operation in the field of peaceful nuclear activities. And my own government, with the help of Franz-Joseph Strauss and the Social Democrat Willy Brand, included at the time, in a governmental note to accompany the Non-Proliferation Treaty, the fact that the construction and the creation of a "Nuclear Superpower Europe" should not be hampered by the Treaty. This means that even now there is the option to combine the British and French nuclear forces and to create a European nuclear body wherein, for example, the Federal Republic of Germany would have the right to determine the use of nuclear weapons.

For many years there was a specific piece of misinformation put about: namely, that plutonium from power reactors was normally not suitable for making bombs. It is not quite clear why so many technically competent people helped to propagate this erroneous notion. Many West European leaders were misled into underrating the dangers of plutonium from power reactors. Professor Albert Wohlstetter's inquiry into this matter helped to put an end to this deception.

The dividing line between the so-called peaceful and destructive uses of nuclear energy was indeed mindlessly weakened. I believe that such a dividing line *never* existed. But why, for example, were attempts made to reprocess spent fuel through the Purex method? The Purex method had been developed to produce especially pure plutonium. For what purpose? To make bombs! The US Atomic Energy Commission distributed the Purex method throughout the world. How can this irresponsible behaviour be explained? Why were there so few warning voices at the time?

The myth had been created that the job of guarding nuclear technology could all be turned over to an international organisation regardless of the type of materials and technical processes involved. Never mind that reprocessed plutonium could rapidly be manufactured into bombs – the Agency in Vienna would safeguard it. Never mind that highly enriched uranium was accumulating in large amounts in many countries – it was under Agency safeguards.

The US Atomic Energy Agency and its successors completely lost track of the whereabouts of hundreds of kilograms of plutonium and highly enriched uranium. It took repeated prodding by the Arms Control and Disarmament Agency to start the job of compiling up-to-date information. And to make matters worse, whenever there occurred a clear violation of the intent of US assistance, as in the case of India's nuclear explosion of 1974; many bureaucrats in Washington would bend over backwards to interpret ambiguities so as to exonerate the foreign government that had defied the intent of Congress.

In a study by Albert Wohlstetter and others (with a foreword by Fred Ikle) it was stated that by 1985 nearly 40 countries would have enough chemically separable plutonium in the spent fuel produced by their electrical power reactors for a few bombs. About half these countries have been planning a capacity by then to separate at least that much plutonium from the spent fuel.

But suppose that a nuclear plant had been operating during Christ's lifetime. Assuming that the operators had stored the radioactive wastes in giant shielded canisters, we would by now have been guarding these wastes for *less than one per cent of the time* that they would have to be isolated from the environment.

The dangers from nuclear power are immediate and lasting. They are awesome and unprecedented. Risk of catastrophic accidents (as at Chernobyl), release of poisonous waste, low-level radiation (even during normal operations), sabotage, the building of atomic bombs by terrorists: these are only some of the terrible possibilities. In spite of these imminent dangers; the nuclear industry argues that we must have hundreds of nuclear plants throughout Europe, to maintain a high living standard. What in fact is happening is that criminal governments and criminal nuclear lobbies are threatening us, the people, by telling us that if we don't go along with what the nuclear establishment advises and wants, we will lose our jobs, our food supply will dwindle, and we will return to the Stone Age. Recently, Helmut Kohl, our Chancellor, explained his arguments for nuclear power exactly along these lines ("Verelendung").

At least countries like Sweden, Holland, Austria, Yugoslavia, Mexico, and the Philippines are beginning to realize that these lies no longer work and are beginning to leave behind their nuclear plans and nuclear nightmares.

Until now, the electrical company lobbies have promoted unlimited energy consumption. Only in a very few cases have there been real programmes for energy conservation which the utility companies also took seriously. The record of unlimited energy consumption in our countries reveals a most insulting assumption: that society exists merely to serve the economy; rather than the other way round.

In the United Kingdom there are, I believe, 16 nuclear power stations with 36 reactors. These include 11 Magnox stations and 5 advanced gas cooled reactor stations. Furthermore, there are experimental prototype reactors; the steam-generating heavy water reactor and the prototype fast reactor. In 1983, nuclear power supplied 17.6 per cent of the UK's electricity, and the nuclear contribution is expected to rise to about 25 per cent when other nuclear power plants come on stream in the next few years. In terms of total energy use, nuclear power contributes about 6 per cent in Great Britain.

In the Federal Republic of Germany we have been working on many scenarios to do without nuclear energy as soon and as quickly as possible, and this must also be a priority in Great Britain. I believe that nuclear power can be phased out if we begin now without power supplies being threatened, including meeting the high demands in the winter.

There must be a combination of creative electricity conservation measures, refurbishing existing coal-fired stations and making them safer, and using the many renewable soft energy paths that we have available to us. Anyone looking at the costs of electricity generation through nuclear power must realize that is a completely mad financial subvention exercise. Many calculations of nuclear economics specifically exclude the original research and development costs, as well as the nuclear fuel cycle that begins for example, when Britain buys uranium from Canada, the USA, Australia and Namibia.

Also at the beginning of the nuclear fuel cycle, extracting uranium from the homelands of the indigenous peoples such as the American Indians and Australian aborigines has a high human cost. Uranium extraction in the homelands leads to many grave health and environmental problems.

A further cost to be considered is that of decommissioning nuclear power stations and disposing of nuclear waste over many, many years.

For these reasons, the figures the nuclear lobby gives us are false.

We must begin developing renewable energy sources such as sunlight, wind power, biomass, wave power, tide energy, hydroelectricity, and the heat in the ground – geothermal energy. The UK is well-endowed with renewable energy resources. It receives a significant solar contribution during the year. This may be sufficient to meet a large part of the requirements for space and water heating of a well designed house. Its wind regime is among the best in the world, and there is major potential for both wave and tidal energy. There is also potential for biomass, especially from wastes. The energy content of municipal and industrial bio-degradable wastes that are currently thrown away is equivalent to around 7 per cent of primary energy use – more than from nuclear power. And there is also energy available in agricultural and forestry residues. A large potential lies in the rock beneath the land mass: geothermal heat.

But as in the FGR, to date UK government support for renewable energies is very limited and very modest compared to that extended to other high technologies. I have recently learnt that in 1984 fast breeder nuclear reactor and fusion research in the UK received £124 million and £30 million respectively, compared with only about £18million for all of the renewable energies. We must have the political will to move away from nuclear power.

I now turn to the topic of radiation. Of all the creatures on Earth, human beings have been found to be most susceptible to the carcinogenic effects of radiation. Cancer is now killing every fourth person in Germany. It is estimated that one in three Americans or Europeans now living will contract the disease at some point.

In addition to giving rise to cancer, radiation also causes genetic mutations (sudden changes in the characteristics of an organism). In 1927, Dr. H.J. Muller was awarded the Nobel Prize for his discovery that X-radiation causes an increase in the number of such mutations in fruit flies. Irradiation-induced recessive mutations might not make themselves immediately apparent. A child might seem normal but carry the deleterious gene and pass it on to the next generation. Diabetes, muscular dystrophy, haemophilia, certain forms of mental retardation, and cystic fibrosis are among the recessive genetic diseases now known. Radiation can also cause chromosomal breakage in a sperm or egg cell. One of the associated diseases is Down's Syndrome.

We are all constantly exposed to some radiation in the form of the natural "background" radiation to which the Earth has been subject for millions of years. Background radiation continues to affect us. This radiation is thought to be responsible for a portion of all the cancers and genetic disorders afflicting us today (Dr. Helen Caldicott). Natural background radiation causes us to age gradually, and increasing that background radiation will accelerate the aging process.

We must here talk honestly and openly about human-made radiation to which most of us are also exposed. Human-made radiation can also initiate cancer and genetic mutations. Medical X-rays are the most prevalent source of radiation that we have known to date. Dr. Carl Morgan, a health physicist, estimates that 40 to 50 per cent of all medical X-rays are unnecessary. And we know the most valuable work of Dr. Alice Stewart in this area of research.

Nuclear power production and the process employed in the manufacture and testing of nuclear weapons are the next most prevalent sources of public exposure. These processes result in the manufacture of hundreds of radioactive elements which are starting to contaminate the food chain and are finding their way into rivers, lakes, and oceans, where these radioactive elements are eaten by fish and are incorporated into the biochemical systems, concentrating in their bodies thousands of times. Contaminated water is taken up by grass and other vegetation and again the radioactive elements are concentrated. Cows grazing on contaminated grass further concentrate the radiation and eventually pass the contamination onto us in the form of milk or meat.

Prominent among the radioactive elements in the production of nuclear power are the beta emitters like iodine 131, strontium 90, and caesium 137. Iodine 131 has a half-life of eight days, but it migrates in the blood to the thyroid gland and may cause cancer there 12 to 50 years later. Strontium 90, which chemically resembles calcium, is absorbed into the bone tissue where it may lead to leukemia and osteogenic sarcoma (malignant bone tumor). Caesium 137, with a half-life of 30 years, concentrates in animal muscle and fish. Ingested by humans, it deposits itself in muscles and irradiates nearby organs. Whether natural or human-made, all radiation is dangerous and there is no safe amount of radioactive material or dose of radiation.

Why? I should pose this question to the other side. By virtue of the nature of biological damage done by radiation, it takes only one radioactive atom, one cell, and one gene to initiate the cancer or mutation cycle. Any exposure at all, therefore, constitutes a serious gamble with the mechanisms of life.

The study by the World Health Organisation concerning the Chernobyl reactor accident states very clearly: "... the current assumption is that there is no threshold dose below which the late effects cannot occur ... " (p.15).

Almost all geneticists agree that there is no dose of radiation so low that it produces no mutations at all. The direct relation between cancer and even minute amounts of radiation has been best demonstrated by Dr. Alice Stewart, who found that only one diagnostic X-ray through the pregnant abdomen increases the risk of leukemia in the offspring by 40 per cent. Every medical textbook dealing with the effects of radiation warns that there is no safe level of exposure. Nevertheless, the nuclear industry and government regulatory agencies, which act in a criminal fashion, have established what they claim to be safe doses for workers and the general public, drawing support from some scientists who believe that there is a threshold below which low doses of ionizing radiation may be harmless.

The International Commission on Radiological Protection (ICRP) originally proposed "allowable" levels of exposure for use by the industry, but not without conceding that these may not be truly safe. It admittedly accorded priority to the promotion of nuclear power. The ICRP noted in its 1966 recommendation (Document no.2): "This limitation necessarily involves a compromise between deleterious effects and social benefits ... it is felt that this level provides reasonable latitude for the expansion of atomic energy programmes in the foreseeable future. It should be emphasized that the limit may not in fact represent the proper balance between possible harm and probable benefit."

This type of philosophy is turning us all into guinea pigs, in an experiment to determine how much radioactive material can be released into the environment before major epidemics of cancer, leukemia, and genetic abnormalities take their toll.

When investigations of low-dose, ionizing radiation revealed that levels of radiation lower than those permitted were causing cancer, US government agencies, for example, attempted to suppress the findings. One such case is the study conducted by Dr. Thomas Mancuso. Its purpose was to determine whether low-level radiation induced any biological effects in nuclear workers at two of the oldest and largest atomic reactors in the USA (Hanford and Oak Ridge). The increasing radiation exposure of workers and the general public by nuclear industries implies tragedy for many human beings. The criminal scandals of Sellafield and Dounreay in Great Britain make clear what we are facing.

And while we worry about the nuclear power industry, the bio-engineering and genetic industries are doing their very best; and I say this in an ironic way, in the future to try to repair those human beings with genetic damage caused by radiation. The atomic state and the gene state: we reject them both!

We must not forget that the latency period of cancer is often 10 to 40 years, and that genetic mutations often do not manifest themselves for generations. We have barely begun to experience the effects radiation can have upon us. The effects of radioactivity on us, our children, and our planet will be irreversible, and thus we must take decisive action now.

It is a cycle of death: the mining of uranium ore, milling and enrichment of uranium, fuel fabrication, the operation of nuclear reactors, reprocessing plants, fast breeders, nuclear sewage, and decommissioning nuclear power plants. Every nuclear power plant will end up on the radioactive garbage heap, because the plant can operate for only 20 to 30 years before it becomes too radioactive to repair or maintain. The costs of decommissioning nuclear power plants are catastrophically high, and these projects are taking two to three years to complete. And what to do with the lethal legacy of nuclear sewage? Nuclear industry projections anticipate a total of 152 million gallons of high-level waste by the year 2000. The cost of preparing even our present load of 83 million gallons for geological disposal is estimated at $20 billion.

Even if there are unbreakable, corrosion-resistant containers, *even if* there are storage sites on Earth, they will have to be kept under constant surveillance by incorruptible guards, administered by moral politicians, living in a stable and warless society, and left undisturbed by earthquakes, natural disasters, or other acts of God for no less than thousands of years (Helen Caldicott).

Seaborg, the discoverer of isotopes, has estimated that 1.6 million pounds of plutonium will be produced by the year 2000. Some of it, perhaps two per cent or more, cannot be accounted for and presumably escapes into the ecosphere during reprocessing, transportation, and other activities. If just two per cent were to contaminate the environment then John Gofman has stated, "Assuredly we can give up on the future of humans."

The National Centre for Atmospheric Research in Colorado revealed in 1975 that more than five metric tonnes of plutonium were thinly dispersed over the Earth as a result of nuclear bomb testing, satellite re-entries and burn-ups, effluents from nuclear reprocessing plants, accidental fires explosions, and leakages. And to think what would happen if SDI ever came to work and nuclear missiles were exploded high above us. The amount of radiation from the destruction of these nuclear weapons would be catastrophic. Whether these missiles land on us or are exploded in the atmosphere high above us – what's the difference?

The vision of nuclear power in its civilian form as a reliable energy source is fading fast. In 1965 the US government predicted that 1000 nuclear reactors would be operating by the year 2000. But new power plant orders have decreased dramatically to none at all in 1977 in the USA.

At this point I would like to mention the case of Karen Silkwood, a laboratory technician at an Oklahoma plutonium factory operated by the Kerr-McGee nuclear company. She collected documents about the hazards and wrong-doings inside the facility. She was on her way to deliver her Documents to a reporter from the *New York Times,* but never arrived. Her car mysteriously went off the road and she was killed. The documents vanished. The question of who killed Karen Silkwood remains an open one. But perhaps not so open at all.

In a momentous decision, a federal jury in Oklahoma found that the giant Kerr-McGee Corporation was negligent in the radioactive contamination of the late Karen Silkwood, and ordered the energy company to pay $10.5 million in punitive damages, and $500,000 for actual damages. Karen Silkwood, the victims of Three Mile Island, Sellafield, Dounreay, the many victims of radiation, whether in Canada, in the homelands of indigenous peoples in the United States or Australia, in Nevada, in the Pacific, or in the Soviet Union; they are all symbols of the unacceptable risk posed by nuclear facilities.

We must also think about the type of society we are becoming through trying to keep nuclear power "safe" and "contained". Nuclear power demands a police state. Some countries experience police states much more than us; for example, those in Eastern Europe. But despite much more freedom and much more democracy than in the countries of Eastern Europe, we experience nuclear power with its accompanying barbed wire, police surveillance, surveillance of workers and their families, private police forces and increasing violence and counter-violence in recent demonstrations. Nuclear power is not at all compatible with a true democratic and decentralized political system.

The British nuclear development programme is said to double Britain's capacity, while questions about safety, health, and economic and political factors remain unanswered. What has happened at Sellafield with leukemia clusters amongst children living near the plant and across the Irish Sea constitutes a grave warning, even without Chernobyl.

Why Chernobyl Still Matters

Rosalie Bertell

Dr Bertell is President Emerita of the International Institute of Concern for Public Health and Member of the Board of Regents, International Association of Humanitarian Medicine. Her most recent book is Planet Earth: The Latest Weapon of War (The Women's Press).

This text was published in The Spokesman 78: Confessions of a Terrorist *(2003)*

Journalists and mathematicians have a way of focusing on one aspect of a complex situation in order to give a snapshot view of its magnitude. For example, one might read in the newspaper that a 'six alarm fire' had occurred in some neighbourhood. This immediately conjures up the image of a very large fire requiring six fire stations to send trucks to the scene. It gives one no clue as to the magnitude of loss of life or property, the water or smoke damage, the impact on human lives and health, ecological impact, and so on. Another example is that of a television show rating scale. If you see an estimate of five million viewers of some special event television, you immediately understand that this is a 'rounded number' meant for comparison only, which does not reveal how many people actually watched the show. Certainly some televisions played to an empty room and some to a large number of people watching the display in the local pub. It gives no indication of whether the watchers reacted positively or negatively to the programme. If the event is important, we expect professionals to fill in the details later.

Another misleading human custom is presenting an event as 'small' when there exist more traumatic forms of the event. For example, the radiation exposure to depleted uranium in the Gulf War is presented as 'small' in the face of a nuclear holocaust. Such exposure is not 'small' for the victims.

Unfortunately, many government officials, physicists, and engineers have used this tactic to deliberately minimise the health effects of radiation, and in particular the immense suffering after the 1986 Chernobyl disaster. For example, some people actually believe that the magnitude of a nuclear accident can be gauged by the potential number of cancer deaths it will cause, and further, that cancer death is the only consequence! Minimalist reporting occurred after the Three Mile Island accident, downwind of nuclear weapon testing, and at serious military accidents like the one which spread plutonium in farmland in Spain. Most recently it has attempted to deny that exposure to depleted uranium weapons has caused severe health damage to the military veterans and the civilians in Iraq, Kosovo and, most likely, in Afghanistan.

The minimalist reporting went even further with Chernobyl. The IAEA (International Atomic Energy Agency) and UNSCEAR (United Nations Scientific Committee on Atomic Radiation) recent statement that 'only 32 deaths occurred, 200 were heavily irradiated and 2000 avoidable thyroid cancers' resulted from the Chernobyl disaster goes well beyond a mathematical short hand which gives an immediate sketch about a disaster. This fifteen-years-later report about a complex, painful situation should be much more precise and believable! It rather tries to obliterate from people's minds and concerns the suffering of millions of persons in rural and un-evacuated areas who were exposed, and hundreds of thousands evacuated but not medically examined victims. When one probes a little more deeply, one finds that the honest scientists and physicians, trying to explain the widespread injuries and long term effects of nuclear exposure, have been silenced.

In fact immediately after the disaster of April 26, 1986, due to International Atomic Energy Agency policy, unless a person had been declared 'overexposed' at the medical tent set up for the 'liquidators' of the disaster, he or she was officially considered to be a 'radio-phobia' case, a purely psychological phenomenon. Local physicians told people that there would be no medical effects of exposure until, perhaps in ten or twenty years, they may happen to develop cancer. But, not to worry! These future radio-genic cancers would be indistinguishable from 'natural' cancers. The physicians soon learned from direct evidence of pathological injuries that this information from the physicists was less than candid. It was not surprising to learn that those who tried to minimise the disaster were the same people charged with promoting nuclear industries, for example, marketing nuclear reactors to the developing nations.

The experience of Chernobyl is not unique, but follows the secrecy pattern used at many lesser accidents which were mishandled in the same way. This has occurred both in the developed and developing world. In particular, I would note the radioactive pollution of the Mitsubishi Asian Rare Earth facility in Bukit Merah, Malaysia, the radioactive waste dumped in Nigeria, and the contaminated food distributed to Egypt, Papua New Guinea, India and other countries during the Chernobyl disaster clean-up.

However, the health problems due to Chernobyl continue to be very acute right now, and demand international attention and action. Scientists and physicians are deprived of their freedom, and the people, especially the children, are suffering. This crisis can serve to point out the serious secrecy, vested interest and collusion of international agencies protecting nuclear technologies. The public face of the nuclear industry has been 'clean and safe'. It is important to unmask this public face, serving as a warning to economically developing countries deciding on energy technologies and bringing needed humanitarian aid to the victims. Preserving the false image of nuclear technology keeps the industry and nuclear agencies in business.

Lessons from Hiroshima and Nagasaki

Unlike the general study of toxic materials, handled by toxicologists, the field of radiation and health has been dominated by physicists, engineers and mathematicians since the dawn of the nuclear era in 1943. Their health related communications differ radically in content from similar communications of health professionals in Toxicology, Occupational or Public Health.

This field of radiation health was, with a few exceptions, taken over by the physicists of the Manhattan Project after World War Two, in their effort to contain the secrets of the nuclear age. Radiation was an effect of the atomic bomb. Secrecy caused these 'hard scientists' to fail to consider the broad range of responses and varieties of vulnerabilities possessed by a living population exposed to this hazard. Such variation in biological responses would have been expected by health professionals.

Because of Hiroshima and Nagasaki, most people now know about acute radiation exposure syndrome, with vomiting, hair falling out, alterations in blood cells, and so on, and this bit of information has been translated into a naïve belief on the part of the public, that unless acute radiation sickness has been documented (often by the government physicists) any subsequent severe illness observed in radiation exposed persons is due to something, anything, but not radiation exposure. This has some historical validity, but at Chernobyl with millions of exposed persons in rural un-evacuated areas, hundreds of thousands evacuated but not medically examined, and with the population's continuous ingestion of contaminated foods for the past fifteen years, demanding documentation of radiation sickness is ridiculous. Even in the Japanese cities radiation sickness went undocumented for many victims. Radiation injury is not predicated on documentation of acute radiation sickness, but rather on the alteration of a cell leading to a fatal cancer. It is well documented the these cellular level events can occur well below the level of exposure which causes overt sickness. The amount of energy released by just one nuclear transformation of one atom of a radioactive material is measured in thousands or millions of electron volts. It requires only 6 to 10 electron volts to break the molecular bounds in the cellular DNA and RNA which carry the genes for life.

In Hiroshima and Nagasaki (1945), exposure and subsequent health records were not complete. The research stations did not begin to select a study population until after the 1950 Japanese census identified survivors and a 1967 dose estimate was derived by the scientists at Oak Ridge National Laboratory in the United States. Deaths prior to 1950 were ignored. Death certificates, which were at times incomplete, were used to determine first cause of death of the study population. Cancers which were not fatal were not reported until 1994. Most survivors are still alive so their 'cause of death' has not yet been studied. Other non-cancer health problems were considered to be 'not of concern' and have not been systematically reported.

There were persons who entered the contaminated territories of Hiroshima and Nagasaki after the fire died down, or who consumed radioactive contaminated food and water, who experienced radiation sickness, but were not officially recognised as 'exposed'. They are in the radiation exposure control group. This is easily explained to the mathematician, who is told that the Hiroshima and Nagasaki studies looked for the effects of the immediate penetrating radiation from the exploding bomb on the persons who were within three kilometres of the hypocentre at that moment. For the military person looking for information on the health effects of radiation due to the bomb, this artificial limitation made some sense. However, if a civil society is seeking information on the effects of man-made radiation on the human body, then all sources of that man-made radiation, including that from nuclear fall-out, food and water contamination, residual radioactive debris at the bomb site, and so on, is important. Changing the definition of 'exposed to man-made radiation' to mean 'exposed to the bomb', and then using this research to back public and occupational health policy is problematic to say the least!

Because of this concentration on the first flash of the atomic bomb, serious mistakes have been made by the radiation physicists in estimating the biological damage done by ingested or inhaled radioactive particles, many of which remain in the body for a long time and even enter into biochemical reactions of the cell's genetic material.

It is this atomic bomb study which appears to be dictating much of the inappropriate behaviour of officials with respect to the medical treatment of survivors of Chernobyl and other nuclear accidents. It has also caused harsh treatment of the honest scientists and physicians who spoke directly for the needs of the exposed suffering people. Many of these scientists and physicians, now in prison or effectively silenced, have conducted well designed and executed scientific studies.

Due to the complications generated by the study of external irradiation by a bomb being used to evaluate civilian exposures to inhaled or ingested radioactivity, and the use of this research to educate young physicists and nuclear engineers, many scientific blunders and administrative problems were generated. The failure to deal with the whole breadth of radiation problems became entrenched in the very agencies which were created in the 1950s to protect the public at risk from atmospheric nuclear testing. I will try to unravel the problems with the International Atomic Energy Agency (IAEA), the United Nations Scientific Committee on Atomic Radiation (UNSCEAR), the International Commission on Radiological Protection (ICRP), the US National Academy of Science Biological Effects of Ionising Radiation Committee (BEIR) and the World Health Organisation (WHO). All of these organisations, except the World Health Organisation, which was relegated to treating the victims rather than understanding the problem, play key parts with respect to current radiation and public health policies and understandings. Ironically, the World Health Organisation, created by the United Nations in 1948, was not given any role in the health assessment of this global threat to human and ecological health.

United Nations Initiatives

Nuclear bombs were first used in war in 1945, when the United States used them against Japan in Hiroshima and Nagasaki. As early as 1946, the United States began atmospheric testing of nuclear bombs in the Marshall Islands, in the Pacific Ocean. The former Soviet Union demonstrated that it had the nuclear bomb in 1949, and there was tangible fear of a nuclear exchange during the Korean War. The United Kingdom began nuclear weapon testing off the coast of Australia in the 1950s, and then on the continent itself and in the Pacific Islands. The first atomic bombs were based on fission, and because of this they were limited in their destructive power. The force of the explosion blew apart the fissioning materials, terminating the explosive energy release. In 1954, the United States tested a thermonuclear device (hydrogen bomb), called Bravo, at Bikini Atoll in the Marshall Islands, demonstrating that a nuclear device with unlimited power could be built. This one was about one thousand times more powerful than the Hiroshima bomb. It was this military accomplishment which prompted the 'Peaceful Atom' speech of President Dwight Eisenhower before the United Nations, also in 1954.

The speech followed a shift in United States Military Policy to dependence on nuclear bombs and a rush towards production of uranium and the technology necessary to carry this out through a major weapon replacement programme: uranium mining and milling, uranium processing facilities, nuclear fuel fabrication facilities, nuclear production reactors, reprocessing facilities and the hazardous transportation and waste associated with each of these industries. In order to obtain American and global co-operation during peace time, there was a perceived need for commercial or so called 'peaceful uses' of nuclear technologies which would justify everyone's co-operation in the nation and the international community. Nuclear electrical production was billed as capable of fulfilling all of the energy needs of the developing world, and being 'too cheap to meter'. It was promoted as the hope of preventing future wars since no country would be in need!

In 1955, the United Nations responded by creating the United Nations Scientific Committee on Atomic Radiation (Res 913(X) 1955) to 'assess and report levels and effects of exposure to ionising radiation'. According to the UNSCEAR website, 'governments and organisations throughout the world rely on the Committee's estimates as the scientific basis for evaluating radiation risk, establishing radiation protection and safety standards, and regulating radiation exposure.' UNSCEAR was envisioned as an organisation of physicists, who at that time were the only ones who could measure radiation since it escapes our senses and requires specialised instruments for detection. They were the experts on the hazard of ionising radiation, but failed to have the expertise to predict the varied human response to exposure to this hazard. In an odd way, perhaps because of their training in physics, they managed to average all exposures over the entire population of the world, now some six billion people. Natural background, because it is ubiquitous, rather homogeneously exposes everyone. However, a localised accident or relatively small workforce's exposure, when averaged over the whole population, can be made to seem trivial. It is not trivial to those who receive the exposure!

The United Nations Scientific Committee on Atomic Radiation became primarily a reporting agency, detailing the measurement of radioactive fallout, worker exposures and eventually emissions from nuclear power plants. I would assume that legislators saw this agency as providing independent monitoring of nuclear activities as a check on predicted pollution and theoretical estimates of harm. Unfortunately, UNSCEAR incorporated into its midst those same scientists who were making the predictions and estimating 'no harm from low level radiation'. No other industry is allowed to monitor itself. We do not ask the tobacco companies to tell us about tobacco's harm, or the pesticide companies to tell us the effects of their products on children. More on this point later.

In 1957, in response to Eisenhower's 'Peaceful Atom' speech, the United Nations also established the International Atomic Energy Agency, which describes itself as 'an independent intergovernmental, science and technology based organisation, in the United Nations family, that serves as the global focus point for nuclear co-operation.' Its mandate is described as: 'to promote peaceful uses of nuclear technology, develop safety standards, and verify that nuclear weapon technology did not spread horizontally to the non-nuclear Nations'. They had no mandate with respect to the nuclear weapons of the five nuclear states. Because of their nuclear watch-dog task, the International Atomic Energy Agency reports directly to the United Nations Security Council.

Response of the World Health Organisation

In 1957, the World Health Organisation, which was founded by the United Nations in 1948, became alarmed about the atmospheric nuclear testing and the proposed expansion of this technology for 'peaceful uses'. It called together eminent geneticists to consider the threat this exposure would pose to the human and ecological gene pool. Professor Hermann Muller, the geneticist who, in 1944, received a Nobel Prize for his work on genetic mutations of the fruit fly using ionising radiation, was a participant at this conference. Although the United States had not sent him as its delegate, he received a standing ovation at the conference for his work, and he consistently opposed the extension of nuclear technology into civilian uses. The conclusion of this expert group was that there was not enough information available in the scientific community to assure the integrity of future generations should the burden of ionising radiation exposure be increased. They called for extreme caution and further genetic investigations, especially in Kerala, India, where there is a high natural background level of radiation, and people have lived in this environment for hundreds of years. These recommendations were never implemented by governments anxious to get on with nuclear activities.

Later, an independent non-governmental organisation in India studied genetic damage in the high radiation background area and found it indeed significantly increased. An article by B.A. Bridges in *Radiation Research* (Vol 156, 631-641; 2001) suggests that genetic mutations due to radiation imply that 'the nature of the radiation dose response cannot be assumed'. There is more complexity than was expected in the health consequences of changed DNA sequences. The serious implications of nuclear pollution for future generations is still an area of research demanding more than ordinary caution.

One can guess at the politics behind a second World Health Organisation conference of psychiatrists, called later in 1957 to consider the Public Health impact of peaceful nuclear activities. These professionals concluded that such activities could cause undue stress to the population because of the association with the atomic bomb. One finds that this has become a mantra for the physicists who have subsequently controlled all information relative to the health impact of nuclear technologies. Most recently, when the United Nations Scientific Committee on Atomic Radiation released its 15 year assessment of the Chernobyl disaster, one of its spokespersons, Dr. Neil Wald, Professor of Occupational and Environmental Health at the University of Pittsburgh School of Public Health, stated: 'It is important that public misperceptions be reduced as much as possible in this area, because unwarranted perception and fear of harm can itself produce avoidable health problems, as well as erroneous societal benefit versus risk judgements.' Loosely translated, Dr. Wald appears to be saying: 'if the public gets upset we will not be able to make our money with this nuclear technology'.

After the Three Mile Island accident in 1979, in response to the people's demand for a health study, the government organised a study headed by a psychiatrist from the Annapolis Naval Academy. He drew concentric circles around the failed nuclear reactor and compared the cancer rates and also the levels of fear and tension of those living within these layers. A sensible study would have looked down wind for air borne radionuclide effects, and down stream for the water-borne effects. This official study found only fear, which was positively correlated with distance from the plant.

There were about 2000 injury cases from the Three Mile Island population taken to court for compensation of health damage due to the radiation exposure. The nuclear company fought all the way to the United States Supreme Court against the courts even hearing these cases, and lost. Then the industry found an old law stating that an expert witness must use the methodology used by other professionals in their field, and using this, the nuclear company managed to disqualify every expert witness (physicians, epidemiologists, botanists, biologists) brought in by the victims. The physicists and engineers claimed sole expertise in the area of radiation health effects. All cases were dismissed by the court without one being heard.

A Deal Between the World Health Organisation and the International Atomic Energy Agency

This potential conflict between those who wished to exploit the new nuclear technology for both profit and military power, and the custodians of the public health, was superficially resolved by an Agreement (Res. WHA 12-40, 28 May 1959) stating that the International Atomic Energy Agency and the World Health Organisation recognise that 'the IAEA has the primary responsibility for encouraging, assisting and co-ordinating research on, and development and practical applications of atomic energy for peaceful uses throughout the world without prejudice to the right of the WHO to concern itself with promoting, developing, assisting and co-ordinating international health work, including research, in all its aspects.' If the reader is confused, so is the writer. To understand this, one needs to know that the health effects of radiation were classified as secret under the United States Atomic Energy Act for national security. The 'international health work' assigned to the World Health Organisation was taking care of the victims. While technically the International Atomic Energy Agency and the World Health Organisation are 'equal' in the United Nations family, those agencies which report directly to the Security Council, as does the Agency, have more status.

In Article I (3) of the WHO/IAEA agreement, it is stated that 'Whenever either organisation proposes to initiate a programme or activity on a subject in which the other organisation has or may have a substantial interest, the first party shall consult with the other with a view to adjusting the matter by mutual consent'. This clause seems to have weakened the World Health Organisation from investigating the Chernobyl disaster, and gave the International Atomic Energy Agency a green light to bring in physicists and medical radiologists to assess the damage relative to their limited knowledge of the health effects of radiation. (Note: while radiologists use ionising radiation in their work, they deal with health damage only after the patient receives therapy levels of radiation.) This first evaluation used a different epidemiological protocol in each geographical area and with different age groups, eliminated all concern for cancers as not having sufficient latency periods and failed to note the extraordinary epidemic of thyroid diseases and cancers. From the point of view of Medical Epidemiology they failed miserably to deal with the reality. The director of this 1991 Epidemiological study, Dr. Fred Mettler, is a Medical Radiologist. There were no Epidemiologists, Public Health professionals or Toxicologists on the International Atomic Energy Agency Team.

The Self-Established
International Commission on Radiological Protection

The United Nations Scientific Committee on Atomic Radiation has continued to be the measurement agency, which verifies that all planned releases of ionising radiation to the environment, and all exposures of workers, are 'acceptable'. It fell to the International Atomic Energy Agency to 'establish or adopt, in collaboration with other competent international bodies, standards of safety for the protection of health and to provide for the application of these standards'.

Neither the International Atomic Energy Agency nor the United Nations Scientific Committee on Atomic Radiation turned to the World Health Organisation to develop such protective health standards. Instead, they both turned to a self-appointed non-governmental organisation formed by the physicists of the Manhattan project together with the Medical Radiologists, who had organised themselves in 1928 to protect themselves and their colleagues from the severe consequences of exposure to medical X-ray. This new organisation, called the International Commission on Radiological Protection (ICRP), has a Main Committee of 13 persons who make all decisions. Members of this Main Committee were originally self-appointed, and have been perpetuated by being proposed by current members and accepted by the current executive committee. No outside agency can place a member on the International Commission on Radiological Protection, not even the World Health Organisation.

The United Nations Scientific Committee on Atomic Radiation 2000 Report was prepared by a Committee including the following seven persons who also serve on the thirteen person Main Committee of the International Commission on Radiological Protection: Prof. Roger Clark (currently the Chair of the International Commission), Prof. Rudolf M.

Alexakhim, Dr. John D. Boice Jr., Prof. Fred A. Mettler Jr. (the same radiologist who headed the International Atomic Energy Agency Chernobyl epidemiological study), Dr. Zi Quiang Pan, and Dr. Yasuhito Sasaki.

It is the International Commission on Radiological Protection which makes recommendations for the protection of human health for workers and the general public. By their own admission, they are not a public or environmental health organisation. They have given themselves the task of recommending a trade-off of predictable health effects of exposure to radiation for the benefits of nuclear activities (including the production and testing of nuclear weapons). Their recommendations were first set in 1957, when the medical radiologists accepted the proposal which had been hammered out by the British, Canadian and American physicists after World War Two.

The original recommendation that workers be allowed 15 rad (150 mSv) per year was opposed by the British National Radiological Protection Board and an independent committee called the BEAR (Biological Effects of Atomic Radiation) funded in the United States by the Rockefeller Foundation. This forced the International Commission on Radiological Protection to reduce their recommendation for nuclear workers to 5 rad (50 mSv) per year. Maximum permissible doses for members of the public were ten times lower. This recommendation remained in effect until 1990, when under pressure from more than 700 scientists and physicians, and after a reassignment of doses at the atomic bomb research centres, the worker exposure was reduced to 2 rad (20 mSv) per year, while exposures to the public were reduced by another factor of five to 0.1 rad (1 mSv) per year.

Who Takes Responsibility?

It is important to note that no agency takes responsibility for these recommendations, and the World Health Organisation is excluded from professional collaboration or comment on them. The International Commission on Radiological Protection recommends, and the Nations are free to implement or not these recommendations. The Nations generally accept International Commission on Radiological Protection recommendations claiming that they do not have the expertise or money to derive their own standards. The recommendations are for a risk benefit trade off, and do not pretend to be based solely (or primarily) on protecting the public or worker health.

The International Atomic Energy Agency states: 'The underlying biological basis of the standards over the last several decades has rested primarily on the United Nations Scientific Committee on Atomic Radiation. This Committee was originally formed during the period of atmospheric weapon testing to assess the physical processes and health effects of fall out, but has since broadened its remit considerably'. UNSCEAR contains and depends on the leaders of the Main Committee of the International Commission on Radiological Protection. Those who set the standards also judge them to be adequate! Usually scientific theory is tested against reality and rejected if it fails to conform. Radiation health predictions are tested against the reality of the victims, and if reality fails to conform to theory, reality is rejected. The suffering is blamed on some unknown cause!

Another body that also assesses radiation risk is the BEIR Committee of the United States National Academy of Science. The BEIR (Biological Effects of Ionising Radiation) Committee was established in the United States around 1978 to counter accusations that the Nevada atmospheric nuclear tests had caused the deaths of thousands of American babies. BEIR is essentially a report and interpretation of the Hiroshima and Nagasaki studies of the effects of the atomic bomb, as previously discussed. These atomic bomb studies do not underpin the radiation standards, which actually were established some 17 years before the 1967 dose assessment for atomic bomb survivors, on which the atomic bomb studies are based, was completed.

The International Atomic Energy Agency radiation standards for nuclear waste were made 'on the basis of recommendations by a number of international bodies, principally the International Commission for Radiological Protection, and estimations of radiation risks made by the United Nations Scientific Committee on Atomic Radiation.' The International Atomic Energy Agency safety requirements for radioactive waste, including standards, codes of practice, regulations, and so on, 'may be adopted by Member States at their own discretion for use nationally'. These Agency requirements are mandatory *only* for the International Atomic Energy Agency itself.

What Happened to the People of Chernobyl?

One can easily imagine that there were civilian victims of radiation sickness in the midst of the chaos during and after the Chernobyl disaster who were never seen at Hospital Six in Moscow. However, the International Atomic Energy Agency continues, even in 2002, to insist that only 32 persons died of radiation exposure at Chernobyl! These 'counted' deaths were all men from the fire fighting brigade identified as seriously exposed and sick by the heroic physicians and other health personnel at the emergency medical tent near the crippled reactor. This type of counting goes even further than the usual mathematical and journalistic approach – it deliberately and maliciously minimises the scale of this disaster and leaves the public vulnerable. Those who were exposed suffer without appropriate medical recognition and help, while those at a distance remain unprepared for another, perhaps worse, disaster.

Moreover, since the land contaminated by the failed reactor was poisoned, the fruits and vegetables grown on it, and the domestic animals who feed on it, and their milk and meat, are also contaminated. Russia, Ukraine and Belarus have taken this contaminated food and, with the advice of the International Atomic Energy Agency, have mixed it with uncontaminated food from other parts of the former Soviet Union. This diluted (or adulterated) food has been given to the people to eat, subjecting them to continuous low doses of internal contamination with radionuclides for the last fifteen years. In Belarus, people actually received money from the government for moving back onto the badly contaminated areas and setting up new farms.

The false claims of the International Atomic Energy Agency have also failed to rally the international community to help the victims of this disaster. People have not responded internationally, with their characteristic generosity, to the tremendous needs of the people whose health and lives were cruelly disrupted. The International Atomic Energy Agency and its companion body, the United Nations Scientific Committee on the Effects of Atomic Radiation, have gone even further in the Spring of 2002, by recommending that Chechen and Central Asian refugees re-populate the still contaminated area around the failed reactor. This raises some very serious questions about the mismanagement of information and communication around this serious disaster.

These two United Nations agencies, namely the International Atomic Energy Agency and the United Nations Scientific Committee on Atomic Radiation, and their partner the International Commission on Radiological Protection, have apparently supplanted the World Health Organisation in speaking to the health risks of this nuclear technology, and in particular, to the post-Chernobyl contamination of the people and the land. Whether or not this land is fit for habitation, or for food production requires health assessment, not a promotional OK from two agencies which have financial ties to the polluting industry!

The World Health Organisation tried to take some initiative on behalf of the suffering people, and in 1996 its Director-General, Dr. Hiroshi Nakajima, organised in Geneva an international conference with 700 scientific experts and physicians, many of whom came from Russia, Belarus and Ukraine. The International Atomic Energy Agency, which to its dismay was not invited to jointly sponsor this international conference, nevertheless blocked publication of the proceedings. The physicians of Chernobyl then organised a conference in Kiev, Ukraine, in June 2001, and invited Dr. Nakajima (who was no longer Director-General of the World Health Organisation) to be their Honorary President. He was asked about the proceedings of the 1996 World Health Organisation Conference about the health of the Chernobyl victims which had never been published. He answered as follows: 'I was the Director-General and I was responsible. But it is mainly my legal department ... Because the International Atomic Energy Agency reports directly to the Security Council of the United Nations ... and we, all specialised organisations, report to the Economic and Social Development Council ... the organisation which reports to the Security Council – not hierarchically, we are all equal – but for atomic affairs ... military use ... and peaceful or civil use ... they have the authority'.

Because of the internal United Nations structure, which is grossly out of date, the voice of the physicians and scientists actually dealing with the situation were not heard. It is outrageous to measure the radiation and then present a theory that no one has been hurt! It is imperative to look at the victims and assess their injury. Internationally, the theoretical voice of the International Commission on Radiological Protection, a non-governmental organisation, which speaks through the International Atomic Energy Agency and the United Nations Scientific Committee on Atomic Radiation, has prevailed. All three agencies have a vested interest in maintaining the reputation of nuclear industries as 'clean and cheap', even if they are not!

The representative of the United Nations Office for Humanitarian Affairs, D. Zupka, was present at the Kiev Conference, and he shared with participants the view of Kofi Annan, who estimated that the number of victims of Chernobyl is nine million. They are predicting that this number will increase. However, their voice is overpowered by the 'scientific' voice of the International Commission for Radiological Protection speaking through the International Atomic Energy Agency and the United Nations Scientific Committee on Atomic Radiation. This seems incredible, but is the heavy burden which we suffer as a legacy of the nuclear secrecy.

Because of the self-serving theoretical predictions and safety recommendations of the International Commission for Radiological Protection which colour the expectations of these radiologists, physicists and engineers, even when they are confronted with the reality of the suffering of the Chernobyl victims, these scientists strongly declare that the observed health problems could not be due to the radiation exposure. Health problems are instead assigned to an unidentified factor in the environment or life-style. Hans Blix, Director of the International Atomic Energy Agency at the time of the Chernobyl disaster, went so far as to say: 'The atomic industry can take catastrophes like Chernobyl every year'. There is an obvious conflict of interest for this agency mandated to promote nuclear technologies!

At the Kiev Conference, Alexey Yablokov, President of the Centre for Political Ecology of the Russian Federation, pointed out that the data used by the United Nations Scientific Committee on Atomic Radiation had been falsified by the State Committee for Statistics, and the officials were arrested in 1999 for this crime. He charged that the United Nations Scientific Committee on Atomic Radiation continued to use this falsified data to support its minimisation of harm.

The medical research of Prof. Y Bandazhevsky, a medical pathologist, Rector of the Medical Institute of Gomel, in Belarus, had to be presented by a colleague, Prof. Michel Fernex. Prof. Bandazhevsky was under house arrest. Belarus received the heaviest fall out from the Chernobyl disaster. After nine years of research in Chernobyl-contaminated territories, he had discovered that caesium 137 incorporated in food, leads to destruction of those vital organs where the caesium 137 concentrates at higher than average body levels. With his wife, a paediatric cardiologist, Bandazhevsky described what he called 'caesium cardiomyopathy', and which others say is a syndrome which will eventually be named after him. The cardiac damage becomes irreversible at a certain level and duration of the caesium intoxication. Sudden death may occur at any age, even in children. After publishing this finding, denouncing government non-intervention policy, and arguing against the lack of resources given to the medical investigation of the disaster, Bandazhevsky was arrested, tried and condemned to prison for eight years.

The trial of Prof. Bandazhevsky was observed by lawyers from the Organisation for Security and Co-operation in Europe (OSCE), from the French Embassy in Minsk, and from Amnesty International. These observers documented irregularities and legal errors from the time of his arrest. In the middle of the night of July 13, 1999, Prof. Bandazhevsky was arrested by a group of police officers, who informed him that the arrest was by presidential decree aimed at fighting terrorism. This was never charged in court. In fact, it was not until four weeks after his arrest, August 1999, that he was finally charged with taking bribes. These proved to be trumped up charges by two defendants who later recanted their testimony saying it was forced under duress and threats. Prof. Bandazhevsky was denied access to a lawyer for the entire duration of his detention, and during the trial there were serious breaches of Belarussian and international law. Amnesty International has listed Prof. Bandazhevsky as a prisoner of conscience. He is not well, and his important research is being kept from his scientific and medical colleagues.

Professor Bandazhevsky is not alone. The Russian, Belarussian, and Ukrainian medical community, though silenced in international circles, was still present and active in alleviating the suffering and noting the causes of their people's pain. Many have carried out detailed high quality scientific studies on the genetic, teratogenic and somatic damage done by radiation exposure. They have confirmed their analyses by demonstrating the effects in animal experiments. The rest of the world is being deprived of this research through heavy handed silencing of the scientists by their national authorities, acting on the recommendations of the International Atomic Energy Agency and the United Nations Scientific Committee on Atomic Radiation, and especially the International Commission on Radiological Protection.

Recommendations

While many individuals have been trying to make known this major United Nations problem, it has been difficult to get this complex situation across to the public in 'sound bites'. Serious study on the part of the United Nations will be needed to undo all of the damage caused. However, it seems possible to make the following recommendations to the United Nations:

- The World Health Organisation should be mandated to review all radiation research and to recommend health-based safety regulations. This mandate should be carried out by health professionals, including epidemiologists, oncologists, occupational and public health specialists, geneticists and paediatricians, (not linked with the nuclear industries or nuclear medicine), rather than other scientists.
- The International Atomic Energy Agency mandate to promote 'peaceful nuclear technologies' should be withdrawn.
- The International Atomic Energy Agency mandate to safeguard the

spread of nuclear weapons should be expanded to include monitoring the reduction and abolition of all nuclear weapons in the nuclear nations.
• The United Nations Scientific Committee on Atomic Radiation (UNSCEAR) mandate needs to include the monitoring of increasing levels of background radiation and nuclear emissions from reactors and nuclear accidents. They should not be entrusted with estimating risk, which is the prerogative of the World Health Organisation.
• Decisions relative to the safety of farmland, food and water ingestion and refugee relocation should be entrusted to the World Health Organisation.
• Investigation into the imprisonment of scientists and physicians who have spoken out on behalf of the public health relative to radiation exposure should be undertaken by a special rapporteur of the Human Rights Commission in Geneva.

With grateful acknowledgements to the Journal of Humanitarian Medicine.

Is Nuclear Power Safe?

Christopher Gifford

The author, a Chartered Engineer, worked as HM Inspector of Health and Safety in mining and quarrying for 25 years. His timely new pamphlet, from which this excerpt is taken, is entitled Nuclear Reactors: Do We Need More? *(Spokesman Books).*

This text was published in The Spokesman 91: Haditha Ethics *(2006)*

More than 400 nuclear reactors operate world-wide. The serious incidents that occurred at Windscale in 1957, at Three Mile Island in 1979, and at Chernobyl in 1986 are well known. Other incidents, especially those that occurred during the Cold War, were not then made known to the public. Some remain to be described. Others such as the releases from the Hanford site in Washington State where eight reactors were built to produce military plutonium between 1943 and 1971 have been disclosed in the United States using Freedom of Information Act provisions. In February 1986, 19,000 pages of documents were released on the application of the Hanford Education Action League. They learned that clouds of radioactive iodine, ruthenium, caesium and other elements were released into the atmosphere contaminating people, animals, water and crops for hundreds of miles. Between 1944 and 1956, 530,000 curies of radioactive iodine was released. The Colombia River became grossly contaminated. In 1954, with six reactors on line, 8,000 curies of radioactive material was dumped into the river each day. By comparison, the radioactivity released at Three Mile Island was 15-24 curies of radioactive iodine. From Chernobyl 3 million curies of caesium 137 was released – a total comparable with the fallout from all nuclear weapons tests to date. The estimate of all radionuclides released from Chernobyl is 50 million curies.

During the Sizewell Public Inquiry it was

argued that human factors, not least human error, had been neglected in the Central Electricity Generating Board's estimates of reactor safety. After the completion of the Inquiry, but before the Inspector appointed to conduct the inquiry, Sir Frank Layfield, had written his report, the Chernobyl explosion occurred, and the early accounts of what had happened left no doubt that human error, notably the defeat by management of the built-in safety systems, was one of the causes. Sir Frank Layfield in his report recommended that there should be further study of human factors by the Central Electricity Generating Board and by the Health and Safety Executive (HSE). The study by the Health and Safety Executive was multidisciplinary and involved all the inspectorates and the research division. In my contribution to the HSE study, I stated that what one most needed to know about human error was that one could depend on it. That conclusion was not disputed by my colleagues, but it did not feature strongly in the evidence submitted to the Hinkley Point 'C' Public Inquiry by the Director General of the Health and Safety Executive, Mr J D Rimington. The report, *The Tolerability of Risk from Nuclear Power Stations,* appended to his evidence, modified the probabilistic risk assessment of a major reactor failure of one per million years of reactor operation to one in 100,000 years to take account of human error. My evidence to that inquiry included the following.

> 'The probabilistic risk assessment was based on guesses about human factors by people who had no experience of power station management.
>
> That comparisons of risk with other methods of electricity generation were not made as required by the Health and Safety at Work etc Act.
>
> That the problems of regulating the industry in private ownership were underestimated, and
>
> That major qualitative differences of risk were ignored, e.g., that not all risk takers would be beneficiaries and that waste management problems would remain for longer than civilisation has existed.'

Fortunately, no failure similar to Chernobyl has yet occurred. The estimates of the consequences of the Chernobyl explosions vary widely from 31 proven deaths by A. Gonzales of the UN International Atomic Energy Agency (IAEA) to nine million people affected – an estimate accepted by Kofi Annan, Secretary General of the United Nations. Estimates of future deaths vary from 40,000 (a contemporary Soviet estimate) to 400,000 by a former Manhattan Project scientist – an estimate based on a correlation of radionuclide release and fatal cancers. Which fatal cancer can be attributed to exposure to which radionuclide is a question for which answers are rare. The Russian Academy of Medical Sciences declared that 212,000 people have now died as a direct consequence of Chernobyl.

The town of Pripyat remains uninhabited. One hundred and twenty thousand people were eventually evacuated from the exclusion zone round the plant.

Mikhail Gorbachev became leader of the Soviet Union on 11 March 1985 when he was elected General Secretary of the Communist Party. By November of that year he had met the US President Ronald Reagan and started the negotiations that led to the largest reduction in the world's nuclear weapons ever negotiated. Five months later, the Chernobyl No 4 reactor exploded. In his memoirs he wrote:

> 'Neither the politicians, nor even the scientists and specialists, were prepared to fully grasp what had happened.
>
> The closed nature and secrecy of the nuclear power industry, which was burdened by bureaucracy and monopoly in science, had an extremely bad effect. I spoke of this at a meeting of the Politburo on 3 July 1986: "For thirty years you scientists, specialists and ministers have been telling us that everything was safe. And you think that we will look on you as gods. But now we have ended up with a fiasco. The ministers and scientific centres have been working outside of any controls. Throughout the entire system there has reigned a spirit of servility, fawning, clannishness and persecution of independent thinkers, window dressing, and personal and clan ties between leaders."
>
> The Cold War and the mutual secrecy of the two military alliances had also been a factor. There had been 151 significant radiation leaks at nuclear power stations throughout the world, but almost nothing was known about them or their consequences. Academician V A Legasov said that the likelihood of nuclear accidents was believed to be very small, and that science and technology throughout the world were not particularly prepared for them. Complacency and even flippancy ruled. I still recall what Academician A P Aleksandrov and Ye P Slavsky told the Politburo immediately after the accident. These men had stood at the heart of our nuclear power industry and were its creators – people who were honoured and respected. But what we heard from them were arguments like this; "Nothing terrible has occurred. These things happen at nuclear reactors ..."'

The effects in Britain 20 years after the explosion require the monitoring of sheep reared on 359 upland farms in Wales where rainfall increased the Chernobyl fallout. If when ready for slaughter the animals have radioactivity levels higher than 1000Bq per kilogram (one Becquerel is one atomic disintegration per second) they are deemed unfit for human consumption and the farmer can be compensated. Farmers mitigate the effects by moving the animals to lower pasture or more distant sites where the grazing is less contaminated and where the animals' radioactivity levels can be gradually reduced to below 1000Bq/kg.

It is not only farm animals that eat vegetation grown on contaminated land. All forms of plant and animal life and water supplies can be affected. The effects are not confined to one generation suffering cancers and reduced immunity to other diseases. Exposed persons and their children and their descendants can suffer mutagenic effects, even teratogenic effects – literally 'monstrous' birth deformities.

The fatalities and the ill health resulting from bomb tests and discharges such as Chernobyl are not as large as those attributable to hydrocarbon extraction and its conversion to electricity or its use in transport. The Chinese mining industry, for example, is reported to have more than 3,000 deaths per year from injuries suffered below ground. Such losses were shown to be avoidable in the UK mining industry which once killed 50,000 miners in 50 years but which, in the 1980s, could produce 100,000,000 tons of coal with good management and better technology and deaths in single figures. Global warming probably caused by human activity could entail even greater losses. Action is required by the precautionary principle.

At a conference attended by World Health Organisation (WHO) and International Atomic Energy Agency personnel and others in Kiev in June 2001, there was concern that the proceedings of a similar conference held in 1995 had not been published. The reason was that the World Health Organisation was allowed to publish material on the effects of ionising radiation only with the agreement of the International Atomic Energy Agency and permission had been withheld under the terms of an earlier agreement. Neither organisation had been sufficiently represented in studying the health effects of the Chernobyl explosion, and a recurring concern even in 2001 was that data was not being collected and reported in the Ukraine, in Belarus to the north, and in Russia.

In a film of some of the proceedings of the Kiev conference made by Swiss film makers, now published by the UK Low Level Radiation Campaign, some of the disagreements remain all too visible. Agency officials and others are shown arguing with medical practitioners that

Chernobyl as a hazardous event is over, that there is nothing that need now cause concern. One argues that there is no difference between exposure to external radiation and exposure from ingested and retained sources of radiation. The manner of dispute by some Agency proponents is revealing to those who may not be expert in the field but who detect arrogance, intolerance and a readiness to denigrate an opponent rather than argue a case. One is left with the impression that the dominant position of the International Atomic Energy Agency *vis-à-vis* the World Health Organisation is not justified and should be ended. One can speculate that having a brief to promote nuclear power has affected the culture of the organisation at the expense of its other commitments.

The film includes scenes of a mother and child. The child born long after Chernobyl has a body mass of 8 kg and a total radioactivity of 10,000Bq. (1250Bq/kg). The explanation for such a level of radioactivity is likely to be the ingestion of radioactive food and water and its incorporation into body tissue where the activity continues as internal emitters. Few people know that in 1990 a Department of Health survey found plutonium in the teeth of every teenage child examined in Britain. The survey of 3,300 adolescents showed minute traces of plutonium in amounts correlating with the distance from Sellafield. When I discussed this finding with my doctor she speculated 'How did it get there?'.

Low level radiation and internal emitters

Clusters of possibly radiation-related disease near nuclear installations led to public concern and investigation by the Committee on the Medical Aspects of Radiation in the Environment (COMARE). In its Fourth Report (1996) on a tenfold excess of childhood leukaemia in the vicinity of Sellafield the Committee included

> '... the current best estimates of radiation doses to the Seascale population is far too small to account for the observed cases of leukaemia and non-Hodgkin's lymphoma that have occurred in the young people of the village during the period of time studied.'

A similar conclusion had been reached at Dounreay. In rejecting radiation as a cause the Committee was in need of another explanation and suggested a population mixing hypothesis which posits that childhood leukaemia is a rare response to a common but unidentified infection. No biological mechanism was proposed.

There is widespread agreement among scientists that there is no safe level of radiation. Radiation-dose-to-disease relationships based on cancers in survivors of the Hiroshima and Nagasaki bombs are questionable because of the possible underestimated low level radiation over long periods and dubious control group selection. It seemed to those who suspected that low level radiation from ingested radionuclides was the cause of the Sellafield cluster that rejection on the grounds of low dose did little more than beg the question. There were other objections from those who knew that the extent of illegal discharges into the environment were not known and who envisaged pathways for sea-borne material to return to the atmosphere and to the land.

At Dounreay, again estimating low dose, the COMARE Committee had not been told of the explosion in the Dounreay shaft which discharged unknown quantities of radioactive material over a wide area and their inquiries did not discover it either. One suspects that if they had asked at the nearest pub someone might have told them. The Nuclear Installations Inspectorate had found many irregularities at the plant, even that the licensee, the United Kingdom Atomic Energy Authority, was not in control and lacked expertise and funds after 36 years of virtual self-regulation. The management and monitoring of stored waste was inadequate. The monitoring of personnel was so lax that employees could choose to leave controlled areas without checks and could have taken contamination to their homes.

After consulting the Committee on the Medical Aspects of Radiation in the Environment, the then Environment Minister, Rt Hon Michael Meacher, MP, announced that a working group would be set up with the remit

> 'to consider present risk models for radiation and health that apply to exposure to radiation from internal radionuclides in the light of recent studies and to identify any further research that may be needed.'

Michael Meacher added that

> 'the Committee's review takes account of the views of all parties in the debate on the risks of radiation. It aims to reach agreement where possible. On topics where differences of view remain after its deliberations it will explain the reasons for these and recommend research to try to resolve them. The Committee Examining Radiation Risks from Internal Emitters (CERRIE) will produce a report that is agreed by all its members. The report will not be subject to amendment by COMARE, the Department of Health or DEFRA and will be

published. COMARE will consider the CERRIE report and advise government on it.'

The committee was unable to agree on many issues; for example, about half of the members believed that the Seascale excesses were not linked to radiation and that the population mixing hypothesis was a possible explanation. The committee failed to achieve its remit and the failure is best explained by the letter of resignation of one of the three members of the secretariat. She said that her work had been altered and distributed to members without reference to her and that she and a third member of the secretariat had been excluded with the effect that there was bias in the work of the committee towards the views of the chairman. She saw no prospect of there being an agreed report.

Although at one stage the committee accepted by a 10 to 1 vote to include what amounted to a modified minority report, it rejected it when all its members received letters from the Department for Environment, Food and Rural Affairs' lawyers warning of personal legal liability for any mis-statement of fact. The third member of the secretariat agreed that he had been excluded and that the views of some of the members had been excised from the final report. Michael Meacher, who by then had ceased to be the minister, wrote to his successor asking for an explanation. Two members, Chris Busby from Green Audit, and Richard Bramhall from the Low Level Radiation Campaign, produced a minority report with an introduction by Michael Meacher in which he expressed disappointment that on such an issue as the increase in childhood leukaemia across Europe after Chernobyl the Committee Examining Radiation Risks from Internal Emitters had presented only one side.

Both reports contain valuable references and some conclusions. The CERRIE report concedes that some risks have been underestimated by a factor of ten. The adoption of that conclusion alone and the international agreements to ban sea dumping and liquid discharges to the sea will make the continued operation of the Sellafield plant difficult. The minority report argues for revision of risk factors by two orders of magnitude. It cites many papers from Chernobyl affected areas. One by Professor Yuri Bandazhevsky, a pathologist, Rector of the Medical Institute of Gomel, on the ingestion of radio caesium includes

> 'Clinical checks on children between 1996 and 1999 show that at levels greater than 50Bq/kg there are pathological changes in vital organs and systems – cardiovascular, nervous, endocrine, immune, reproductive, digestive excretory

and eyes. Caesium concentrations in the placenta reveal a relationship with nervous system defects in the foetus. The health condition of the population is a disaster but being a physician myself I cannot accept it as hopeless. With all my faith in God and life I appeal to anyone who can influence it: do your best to improve the situation. There is nothing more precious on this planet than life. And we should do everything possible to protect it.'

He is the author of over 400 publications, a member of five academies and the holder of five international awards. His critics, one from the International Committee on Radiological Protection, explain the phenomena as psychosomatic effects of radiophobia generated by such publicity as his own. He criticised his government for lack of involvement. In 2001 he was arrested, charged with corruption, which he denies, and sentenced by the military court of the Supreme Court of Belarus to eight years imprisonment. He was adopted by Amnesty International as a prisoner of conscience. The European Parliament awarded him the Passport for Liberty and the European Union called for a review of his trial. Data on caesium 137 effects was not included in the conference record.

Was the Chernobyl explosion nuclear?

At the Hinkley Point 'C' Public Inquiry it was suggested by at least one witness that the Chernobyl explosion was a nuclear explosion. The suggestion was vigorously refuted by the Central Electricity Generating Board. One of the design requirements for the licensing of a reactor in the United Kingdom is that the containment shall be capable of withstanding the effects of any fault. Safety Assessment Principle 152 requires 'The containment should adequately contain such radioactive matter as may be released into it as a result of any fault in the reactor.' Clearly if nuclear explosions are possible a licence should not be granted. That they were granted suggests that the Nuclear Installations Inspectorate reject the view that nuclear explosion is possible. What then can we make of the information that two nuclear engineer Fellows of the Royal Society and of the Royal Academy of Engineering have recorded their opinion that the Chernobyl explosion was nuclear? It did displace the 2,000 tonne concrete cap from the reactor.

The Secretariat of the Nuclear Free Local Authorities quote Sir John Hill, a former chairman of the Atomic Energy Authority, who wrote in *ATOM,* the Atomic Energy Authority house journal:

'When the Americans chose graphite moderated water cooled piles for plutonium production they recognised that a failure of the water supply or control system could result in prompt criticality and a *nuclear* explosion such as happened 40 years later at Chernobyl.' (my emphasis).

Jack Harris is a former Central Electricity Generating Board nuclear metallurgist who writes a monthly column in the Journal *Materials World,* one of the journals of the Institute of Materials, Minerals and Mining. An article he wrote in June 2004 makes clear his acceptance of his colleague Ross Hesketh's view that the Chernobyl explosion was a nuclear explosion. Jack Harris, now a university professor, has been a Fellow of the Royal Society (FRS) since 1988. It would be interesting to know how many other Fellows of the Royal Society, Fellows of the Royal Academy of Engineering, nuclear engineers and physicists share the view that our more recently built reactors are capable of blowing themselves to bits. We really ought to know. Perhaps the matter was decided by the Chernobyl experience but was too difficult to contemplate, let alone acknowledge. It could be the best kept nuclear secret since 1986.

The nuclear industry remains uninsurable worldwide. United Kingdom legislation allows the industry to operate with what is obviously inadequate cover and provides for further cover by the government and the taxpayer.

Fukushima
The improbable can happen

Zhores Medvedev

This article was published in *The Spokesman 113: NATO? - No thanks!* (2011).

Dr Medvedev revealed to an unbelieving world the nuclear disaster which took place at Kyshtym in the Urals during the 1950s, confounding nuclear 'experts' who denied his case. He tells this story in a preface to the new edition of his classic study, The Legacy of Chernobyl, *which Spokesman are publishing on 6 August to mark Hiroshima Day. This excerpt is taken from the new preface.*

The first nuclear reactor for the Fukushima I plant was designed in the early 1960s, ordered in 1966, and put into operation in 1971 for the Tokyo Electric Power Company (TEPCO) by General Electric. It was modest in terms of power; a 460 MW boiling water reactor (BWR). The second reactor of the same type was more powerful (784 MW), and the last (the sixth reactor), which came into operation in 1979, was 1100 MW. The design was expected to withstand seismic events of magnitude 7,5. This was the force of the California earthquake of 1952. The San Francisco earthquake of 1906 was 7,8. The Great Tokyo Earthquake of 1923 was magnitude 8,3. The earthquake on 11 March 2011 was magnitude 9.

The Fukushima I nuclear accident is now considered the second largest after Chernobyl. But it is still developing, and might yet take the lead in the IAEA list of nuclear accidents. It is much more complex because it involves several reactors and the spent fuel storage tanks, with about 25 times more radioactivity than there was in the Chernobyl reactor. The picture of the accident grows darker and darker almost daily: partial meltdowns in reactors 1, 2 and 3; hydrogen explosions which destroyed the upper parts of buildings housing the reactors; damage to the containment inside reactor 2; fires and leaks. The amount of radioactivity released into the environment has already reached the Chernobyl level. However, in Chernobyl the release was gases and aerosol into the air. In Japan, there are mostly radioactive solutions which contaminate soil and sea. (In Japan, there is also much more plutonium.)

Igor Ostretsov, a nuclear engineer of Soviet pressurised water reactors, whom I consulted, wrote that the location of the emergency power generators so close to the sea and at sea level, just facing the great tectonic fracture, was a serious mistake. He also considers unfortunate and unsafe the location of very heavy spent fuel storage tanks filled with water in the same building above the reactor. Such a location made it easier to load the fuel rods from the reactor into the storage pool, but it also made 'suspended' storage tanks very vulnerable to any earthquake.

The earthquake damaged these tanks, causing them to leak. Their location made it difficult to refill them with cooling water. Helicopters and fire engines were used out of desperation.

Another design fault identified by Ostretsov was the absence of a ventilation system for radiolytic and zirconium-steam-reaction-produced hydrogen gas. This resulted in an accumulation of the gas in the reactor building, and caused explosions which destroyed the building and many critical systems, particularly the cooling loops. The use of seawater was another desperate measure, as the water evaporates, leaving salt, which further damages the fuel rods in the core and in storage pools.

The boiling water reactor (BWR) system has one more problem. The same water, which functions as a neutron moderator and is part of the fission control, is also feeding steam directly to the turbines without an intervening heat exchanger. This purified and deionised water is pumped to the bottom of the fuel channels and boils, producing steam used to drive the turbines. This water accumulates fission radionuclides.

Heavily contaminated water, particularly with iodine-131 and caesium-137, is the main problem. Reactors do not produce carbon dioxide, which is an advantage. But they produce an enormous amount of heat. Working reactors therefore consume a huge amount of cooling water; 21,000 tons per hour in Fukushima I-1 , 33,300 tons per hour in Fukushima I-3, and 48,300 tons per hour in Fukushima I-6. Even after shutdown, residual heat from accumulated radionuclides constitutes up to five per cent of project power (depending on the fuel cycle), enough to cause meltdown. Partial meltdowns were reported at Fukushima after the emergency shutdown. Thus, several thousand tons of water per hour are still needed to cool the residual heat of the cores and the spent fuel in storage. With the circulating systems damaged, this water has to be dumped in the sea. There is no project provision to store this amount of radioactive water. Temporary storage was possible only for the reduction of iodine-135 (half-life 6,7 hours) and iodine-131 (half-life 8 days). The iodine isotopes were produced in working reactors. (Nearly 80 million curies of radioactive iodine were released into the air in Chernobyl).

Now, three months after the Japan earthquake, the danger from iodine has diminished. The main problems are strontium-90, caesium-137, plutonium and a few more long-lived radionuclides. The danger of new meltdowns is not yet over. The main problem for years to come will be managing more than 500 tons of spent fuel in the reactors and in storage pools, more than 4 tons of which is plutonium. The cooling systems of the reactors and spent fuel tanks were found beyond repair and the current methods of cooling continue to wash out the radioactivity into the environment. The project to dismantle the whole nuclear plant with its six reactors might take many years.

Fukushima's Radioactive Elements

Helen Caldicott

This article was published in *The Spokesman 114: From Hiroshima to Fukushima* (2011).

Dr Caldicott has devoted much of her life to an international campaign to inform the public about the medical hazards of the nuclear age and the necessary changes in human behaviour to stop environmental destruction. Her most recent book is a revised and updated edition of If You Love This Planet *(W.W.Norton & Co.)*

Huge quantities of radioactive elements, more than anyone has been able or willing to measure, have been continuously released into the air and water since the multiple meltdowns at the Fukushima Daiichi Complex in Japan on and around 11 March 2011.

This accident is enormous in its medical implications. It will induce an epidemic of cancer the likes of which the world has only rarely experienced, as people inhale the radioactive elements, eat radioactive vegetables, rice and meat, and drink radioactive milk and teas.

As radiation from ocean contamination bio-accumulates up the food chain, radioactive fish will be caught thousands of miles from Japanese shores. As they are consumed, they will continue the cycle of contamination, proving that no matter where you are, all major nuclear accidents become local.

In 1986, a single meltdown and explosion at Chernobyl covered 40% of the European landmass with radioactive elements. Already, according to a 2010 report published by the New York Academy of Sciences, almost one million people have perished as a direct result of this catastrophe, yet this is just the tip of the iceberg.

There is confusion and misunderstanding in the media, and amongst politicians and the general public, about what nuclear accidents, particularly the accident at Fukushima, will mean medically. It will be useful to explain how radiation induces disease and what sort of radioactive material is contained in a nuclear power plant.

Fact number one
According to every version of the BIER study by the National Academy of Sciences, up to and including the most recent in 2007 – The Biological Effects of Ionizing Radiation No. V11 (BIER VII) – no dose of radiation is safe. Each dose received by the body is cumulative and adds to the risk of developing malignancy or genetic disease.

Fact number two
Children are ten to twenty times more vulnerable to the carcinogenic effects of radiation than adults. Foetuses are thousands of times more so. Immuno-compromised patients, and the elderly, are also extremely sensitive.

Fact number three
Ionizing radiation from radioactive elements, including radiation emitted from X ray machines and CT scanners, damages living cells. This can result in cancer.

How? Simply speaking, there is a gene in every cell called the regulatory gene. It controls the rate of cell division. If this specific DNA sequence is hit by radiation, the cell will either be killed or, alternatively, the regulatory gene can be bio-chemically altered. This is called a mutation. It is impossible to know if this damage has taken place in your body. The cell will sit silently for many years until, one day, instead of dividing in a controlled fashion, by mitosis, to form two daughter cells, it begins to reproduce uncontrollably, producing trillions of cells. That is a cancer. A single mutation in a single gene can kill you. This process is accelerated in children.

Fact number four
The latent period of carcinogenesis
The incubation time for leukaemia is five to ten years, but for solid cancers (such as breast, lung, thyroid, bone, kidney, and brain) the incubation period ranges from 15 to 70 years. All types of cancer can be induced by radiation.

Fact number five
The reproductive cells in the body, the eggs and sperm, are even more important genetically than normal body cells. Each egg and sperm has only half the number of genes as those in a normal cell so that when they unite, a new normal cell is produced which goes on to form an embryo, then a foetus, then a fully formed baby. Every gene in an egg or sperm cell is

precious because these genes control the characteristics of the new individual. Therefore, if normal genes are mutated by radiation the new baby could be born with a genetic disease, or will carry abnormal genes for diseases such as cystic fibrosis and diabetes, or inborn errors of metabolism to be passed on to future offspring. There are over 2,600 genetic diseases now described in the medical literature.

We all carry several hundred genes for genetic disease but, unless we mate with someone carrying the same gene (such as cystic fibrosis), the disease will not become manifest. These abnormal genes have been formed over aeons by background radiation in the environment.

As we increase the level of background radiation in our environment from medical procedures, X-Ray scanning machines at airports, or radioactive materials continually escaping from nuclear reactors and nuclear waste dumps, we will inevitably increase the incidence of cancer as well as the incidence of genetic disease in future generations. Mutated or abnormal genes are passed down from generation to generation in perpetuity.

Fact number six
There are basically five types of ionizing radiation:
1 X-Rays (usually electrically generated), which are non-particulate, and only affect you the instant they pass through your body. You do not become radioactive but your genes may be mutated.
2 Gamma rays, similar to X-Rays, emitted by radioactive materials generated in nuclear reactors and from some naturally occurring radioactive elements in the soil.
3 Alpha radiation, which is particulate, and composed of 2 protons and 2 neutrons, emitted from uranium atoms and from other dangerous elements generated in reactors (such as plutonium, americium, curium, einsteinium, etc – all known as alpha emitters). Alpha particles travel a very short distance in the human body. They cannot even penetrate the layers of dead skin in the epidermis to damage living skin cells. But, if these radioactive elements get into the lung or the liver, bone or other organs, they transfer a large dose of radiation over a long period of time to a very small volume of cells. Most of these cells are killed, but some on the edge of the tiny radiation field will survive. Their genes will be mutated, and cancer may later develop. Alpha emitters are among the most carcinogenic materials known in medicine.
4 Beta radiation, like alpha also particulate, is a charged electron emitted from radioactive elements such as strontium 90, caesium 137, iodine

131, etc. The beta is light in mass, it travels further than an alpha particle but does the same thing, mutates genes.
5 Neutron radiation is released during the fission process in a reactor or a bomb. Reactor 1 at Fukushima is still periodically emitting neutron radiation as sections of the molten core become intermittently critical. Neutrons are large radioactive particles that travel many kilometres, and they pass through everything including concrete, steel etc. There is no way to hide from them and they are extremely mutagenic.

So, let's describe just four of the radioactive elements that are continually being released into the air and water at Fukushima. Remember, though, there are over 100 such elements each with its own characteristics and pathways in the human body. All are invisible, tasteless and odourless.

Caesium 137 is a beta and gamma emitter with a half-life of 30 years. That means in 30 years only half of its radioactive energy has decayed, another 30 years to decay again to half, so it is detectable as a radioactive hazard for some 600 years. For the first 300 years (the standard 10 times the half-life calculation) the levels remain of regulatory concern, but for 300 more years the radiation is still detectable. As there is no safe dose, these levels are still significant and still a hazard. When it lands on the soil it bio-concentrates in grass, fruit and vegetables to many times background levels. It then bio-concentrates tens to thousands of times more, in meat and milk, as animals eat the fruit and vegetation. It concentrates the highest in the human body, the top of the food chain. It is very worrying that it is not, in fact, the adult human body, but that of the newborn infant, which is at the very top of this chain. Because caesium resembles potassium, which is ubiquitous in every cell in our body, it tends to concentrate most highly in brain, muscle, ovary and testicles. There it can cause brain cancer, muscle cancers (rhabdomyosarcomas), ovarian or testicular cancer and, most importantly, mutate genes in the eggs and sperm to cause genetic diseases in future generations.

Strontium 90 is a high-energy beta emitter, half-life 28 years, detectably radioactive for 600 years. As a calcium analogue, it is known as a bone-seeker. It concentrates in the food chain, specifically milk (including breast milk), and is laid down in bones and teeth in the human body, where it can irradiate a bone forming cell, or osteoblast, causing bone cancer; or mutate a white blood cell in the bone marrow which can initiate leukaemia, a cancer of the white blood cells.

Radioactive iodine 131 is a beta and gamma emitter with a half-life of eight days, so it is a hazard for 20 weeks. It bio-concentrates in the food chain, in vegetables and milk, and specifically concentrates in the human thyroid gland where it is a potent carcinogen inducing thyroid disease and thyroid cancer.

Plutonium, one of the most deadly, is an alpha emitter, so toxic that one millionth of a gram will induce cancer if inhaled into the lung. It is transported from the lung by white blood cells, then laid down in thoracic lymph nodes where it can induce Hodgkin's disease or lymphoma. Because it is an iron analogue it combines with the iron transporting protein transferring and concentrates in the liver, a cause of liver cancer; the bone marrow in the haemoglobin molecule, a cause of bone cancer, leukaemia, or multiple myeloma. It concentrates in the testicles and ovaries where it can induce testicular or ovarian cancer, and/or mutate genes to induce genetic disease in future generations. It is one of the few toxic substances that can cross the placental barrier which protects the embryo. Once lodged within the embryo, the alpha particle could kill a cell that would form the left side of the brain, or the right arm – as thalidomide, the morning sickness drug, did years ago.

The half-life of plutonium is 24,400 years, so it can cause harm for 500,000 years; inducing cancers, congenital deformities, and genetic diseases for the rest of time. Not only in humans, but in all life forms.

Plutonium is also fuel for atomic bombs. Five to ten pounds will fuel a weapon which would vaporize a city. Each reactor makes 500 pounds of plutonium a year. It is postulated that one pound of plutonium, if adequately distributed, could kill every person on earth from cancer.

Fact number seven
In summary, the radioactive contamination and fall-out from nuclear power plant accidents will have medical ramifications that will never cease. It will affect future generations, in human terms, forever; inducing epidemics of cancer, leukaemia and genetic disease.

Last thoughts
This is a pivotal time in human history. We watch radiation slowly blanket Japan, a country with four reactors in trouble, in the midst of the worst industrial accident in history, facing an uncertain future of terrible health effects, and catastrophic environmental damage. We watch, helpless, as Fukushima fall-out traverses the Northern Hemisphere, turning up in milk, food, and water; on tourists in airports; and products in shipping bays around the world. We are seeing, and understanding, that all fall-out is local.

There is a reactor in the United States in the middle of the flooding Missouri River, and another just downstream, also in danger should major dams fail. Wildfires recently raged within two miles of the Los Alamos National Laboratory's grounds, a storage place for high and low level nuclear waste from the Cold War, an area where miles and miles of burning land is contaminated by legacy fall-out from atomic testing. Similar wildfires raged over contaminated land in Russia last summer. With ageing nuclear reactors and weapons becoming both more volatile, and more vulnerable, it is time to ask again, this time more forcefully: what is peaceful about nuclear power?

We are staring global warming in the face. Water shortage, famine, rising temperatures, wild weather, and climate refugees in numbers unseen in history are staring back at us. You can't stare down climate change, as the nuclear industry would like to; instead we need to power down our old, wasteful and expensive, dangerous sources of energy and start plugging in to a renewable, sustainable-energy future. We have the money, we have the technology, and we have the time – just barely. If politicians lack the political will, then now is the time for the will of the people to speak louder. There is no other world suitable for life. We either change, or we see the end of this world as we know it.

With grateful acknowledgements to the author for permission to republish. For more discussion of what is happening at Fukushima, and related issues, please see online (www.nuclearfreeplanet.org).

Fukushima's Quagmire

Hachiro Sato

This article was published in *The Spokesman 114: From Hiroshima to Fukushima* (2011).

Mr Sato is a member if Iitate Village Assembly in Fukushima prefecture, which is about 24 miles north-west of the nuclear disaster site.

He gave this address in Hiroshima on 4 August 2011, during the World Conference Against Atomic and Hydrogen Bombs.

Let me begin by expressing our deepest gratitude to you for the warm support you have given us in goods and money, as well as in providing places to stay for the victims of the earthquake and the accident at the Fukushima Daiichi nuclear power plant.

The accident at Fukushima Daiichi is heading further into the quagmire, rather than coming under control. Right after it happened, about 120,000 people from Fukushima took refuge within or outside the prefecture. As of July 15 2011, this number has declined to 76,194, but the number of people who have fled from Fukushima prefecture has risen to 45,242, meaning 2.2 per cent of its population are forced to live elsewhere. One can only imagine the frustration and anxieties they live with every single day, having no idea about when they will be able to go back home. Those who continue to live in Fukushima, in places not designated as evacuation zones, are also deeply concerned: 'How many years can we live in such an environment where low levels of radiation continue to exist? Is it really true that there will be no ill-effects on our children?' No one has a clear-cut answer to these questions, and this is what concerns us most.

Administratively speaking, the village of Iitate, my hometown, became what it is today when it was merged with other neighbouring communities on 30 September 1956. Mountains and forests cover as much as 75 per cent of its 230 square-kilometre land area. It is a genuine farming community where our means of living are the production of rice, livestock, tobacco leaves, vegetables, flowers and ornamental plants. 30.4% of our 6,170 inhabitants are old people, over 65 years old. The average yearly temperature is 10 degrees Celsius and the mean annual rainfall is about 1,300 millimetres. We have a cold climate in which the cold and wet easterly/northeasterly winds drift down during the early summer days. This, coupled with the late frost that comes as late as mid-May, damages our crops. We don't have much snowfall in winter, but the temperature can fall as low as 15 degrees Celsius below zero.

In spite of these unfavourable conditions, our ancestors and we have taken much pains and made great efforts in building the community. Today, Iitate has been chosen as one of 'Japan's most beautiful villages'. This year, 2011, was to have seen a leap of progress with the building of more factories and making local specialties for sale.

On 12 March, hydrogen explosions occurred at the No.1 and No. 3 reactors of the Tokyo Electricity Power Company's (TEPCO) Fukushima Daiichi nuclear power plant, followed later by fires at the No. 4 reactor, and partial destruction of the containment vessel of the No.2 reactor. These critical man-made disasters caused highly radioactive materials to fly in all directions and to spread over vast areas, including our village. All our villagers were forced to evacuate their homes, leaving everything behind including land, property, livelihoods and even hopes.

The need now is for the evacuees to prepare to return to the village and rebuild the infrastructure so that their health and safety will be ensured as well as the right to live. But the government, utility, prefectural and local administrations are too slow in carrying out measures in response to these needs. All this makes things even more difficult, along with the lack of credible information. For four months, since the start of the present disaster, the government and the utility have withheld important information. They have lied about what has happened, and even underestimated the present dangers. In the early stages, we accepted evacuees from Minami-soma City and Futaba Town. We assisted evacuees until March 18. It is regrettable that we supplied water and food contaminated with radioactive materials. On March 22, we learned from media reports that radioactive materials were detected in broccoli. Since then, radiation levels were announced every other day for milk, water and soil. Infants, pregnant women, children, young people and women were advised to leave the village voluntarily. Families were broken up, and people were forced to live in different evacuation centres.

The village authorities did not order all residents to leave their homes to fulfil their administrative responsibility. Instead, they requested the government to take special measures and carry out decontamination of firms and long-term nursing homes, which the village wanted to keep until the government designates their areas as exclusion zones. Many villagers began to voice angrily their rejection of nuclear power plants in their village. They did not know what to do in the present situation. Despite growing uncertainty, village mayor Sugeno Norio submitted a proposal to the central government, saying that the village had no intention of becoming an anti-nuclear flag-bearer.

Eight business establishments and a long-term nursing home continue to operate in the village. More than 800 villagers, including a 370-member 'security team' working three shifts to protect the village and its assets, have returned to the village. This is what's happening in our village. Isn't this the way to turn the people into guinea pigs to study the effects of low-level radiation exposure? We simply want no more nuclear power plants anywhere, whether they are in Fukushima, Japan or elsewhere in the world.

Our village has hosted nuclear power plants in the name of 'national policy' by accepting what the government explained as 'the use of nuclear energy for peaceful purposes'. No one imagined that a level 7 nuclear disaster would wipe out towns and villages, and displace large numbers of people to unknown places, where they would be unable to foresee what would become of them. The accident exposed the brutal outcome of over-optimistic views about Fukushima and on the part of national governments, which have believed the 'safety myth'.

Another factor that delayed an evacuation order was the lack of sufficient radiation monitoring sites in the prefecture, which hosts ten nuclear power plants! Right up to the disaster, there were only 23 monitoring posts within 20 kilometres, and only one near the prefectural government building, all under the control of the Local Development and Promotion Bureau. Regarding the prompt distribution of accurate information, what TEPCO and the Ministry of Trade, Economy and Industry are doing has been contrary to that which is needed. They have withheld needed information. Their underestimation of the crisis appears to be intentional. Fukushima residents are sick and tired of the comments they continue to make on television. To date, they refuse to admit that the nuclear accident is man-made.

In areas that are not designated as exclusion zones, parents are desperate to protect their children from radiation exposure. They are not sure if the one millisievert per year safety limit is really appropriate for the protection of children's health. They are confused by mixed and various kinds of information they see on the internet. We must seek to use the power of collective wisdom from around the world in order to save our children. To this end, we call for an authoritative independent body to be established.

The Fukushima Joint Centre for Post-disaster Reconstruction, which Fukushima Gensuikyo is working with, conducted an opinion survey on nuclear power plants. It shows 84.5 per cent of respondents favouring the decommissioning of all nuclear plants in Fukushima, and 79.3 per cent wanted TEPCO to compensate for the damage. One respondent said:

> 'The Fukushima Prefectural Government shouldn't allow TEPCO to take charge of disaster management at its power plant because it's just like asking a robber to look after your house.'

Another respondent said,

> 'Do you understand how painful it is to hear your child say he would die of cancer? We want safety to be restored in Fukushima!'

One more said:

> 'We need absolutely safe energy sources in place at any cost.'

Outrage and demands spilled out of the respondents' comments.

Following the nuclear disaster at Fukushima Daiichi, Italy and Germany decided to shut down their power plants. In Fukushima, the governor, Sato Yuhei, publicly reversed his position in the Fukushima Prefectural Assembly session of June 27. He said that Fukushima Prefecture should seek to build a community that does not depend on nuclear energy.

The Fukushima Prefectural Liaison Council for the Safety of Nuclear Power Plants, which consists of the prefecture's democratic organizations, has, for 38 years, been active in pointing out the danger of nuclear power plants, but neither the Fukushima prefectural government nor TEPCO has listened. They should have realized the danger before we had to go through such ordeals. Living in the only country to have been attacked with nuclear weapons, we should also have realized the precariousness of so-called 'atoms for peace'. To begin with, was it appropriate to regard what used to be weapons as safe? The fact is that we have no scientific method for the safe storage of spent fuels. Clearly, humankind cannot coexist with what is 'nuclear'. I believe it is our duty to work for the swift development of renewable energy sources in order to leave a safe and sound world for our children. We must call for no more nuclear victims.

Nuclear Explosions

On not learning the lessons of Fukushima and Chernobyl

Christopher Gifford

This article was published in *The Spokesman 115: Syria and Iran* (2012).

The author is a Chartered Mining Engineer. While in the Health and Safety Executive, he worked with nuclear installations inspectors and others on risk assessments in high-risk industries.

The Earthquake
On 11 March 2011 at 14-46 local time, a magnitude 9 earthquake occurred under the Pacific Ocean 110 miles east north-east of Fukushima on the east coast of Japan. Fukushima Dai-ichi, (Fukushima 1) is a nuclear power station complex of six boiling water reactors.

Reactors numbered 1, 2 and 3 were generating steam for electricity power production. Reactor No 4 was under maintenance and its nuclear fuel had been removed to storage ponds. Reactors 5 and 6 were shut down for maintenance.

HM Chief Inspector's Report
This account of the events at Fukushima 1 is based largely on the Interim and Final Reports by Dr Mike Weightman, HM Chief Inspector of Nuclear Installations in the United Kingdom. He was asked to provide the reports by the Secretary of State for Energy and Climate Change. The British government was monitoring the events for the safety of British nationals in Japan, and for the international effects of the disaster, now ranked 7 on the International Nuclear and Radiological Event Scale (INES) (the highest level).

Dr Weightman was assisted by a Technical Advisory Panel and respondents to his Interim Report of May 2011. He led a mission of experts in Japan and visited Fukushima and several other nuclear sites. His Final Report also acknowledges the assistance of the report of the Japanese Government to the International Atomic Energy Authority (IAEA) Ministerial Conference on Nuclear Safety, June 2011, and other sources, while explaining that, because of the effects of the tsunami, not everything is known about the disaster and may never be known. Other work is under way or planned which also seeks to learn lessons from the disaster, such as the European Council 'Stress Tests' on nuclear installations (to find weaknesses in existing reactors and methods) and the work of the Nuclear Energy Agency (NEA) of the Organisation for Economic Co-operation and Development (OECD) and the IAEA on which a report, for publication in October 2012, is planned.

Electricity grid failure
The Fukushima 1 Reactors 1, 2 and 3 were protected by seismic sensors which detected shocks of 0.56g peak horizontal acceleration and triggered safe shutdowns. Similar shocks had been recorded on at least one other occasion and the reactor buildings, as before, survived without collapse, although some damage to equipment was expected. The earthquake caused electricity grid pylons to collapse, some affected by landslip, and all standby incoming power supplies to the power station failed. All but one of 13 standby diesel generators survived the shocks and were started to run pumps for water to cool the reactors and the spent fuel ponds and to provided lighting and other services.

The tsunami
In less than one hour after the earthquake, a 14m tsunami wave inundated the complex to a depth of six metres and caused extensive damaged with much debris. Fourteen other nuclear reactors on the east coast of Japan were also affected. The Fukushima 1 diesel generators and switchgear were unusable. Control and instrumentation equipment and lighting were also affected. The engineers operating the reactors lacked information about the status of the reactors and spent fuel storage ponds, and their efforts to provide and maintain emergency cooling of both required much improvisation and exposure to hazards.

Reactor explosions
Within the next two days, explosions occurred in the buildings housing reactors 1, 2, 3 and 4 and some secondary containment was destroyed. The explosions were probably of hydrogen gas produced by zirconium alloy fuel cladding reacting with steam. Fuel in the reactors is believed to have melted and in some may have breached the primary containment. There were major releases of radioactivity, initially to air but later by leakage or discharge of contaminated water to sea. In consequence of the releases to air, eighty thousand people were evacuated from their homes with little

prospect of a return other than to collect any property that they could carry in a visit of less than 2 hours.

On 29 October 2011 *The Guardian* newspaper reported on a paper published online by the journal of Atmospheric Chemistry and Physics in which European and US experts estimated that the release of Caesium 137 from Fukushima was 42% of that from Chernobyl and more than twice that from Fukushima as reported by the Japanese government.

How much of the Japanese nuclear energy industry will survive?

After the tsunami only 19 of Japan's 54 reactors continued to operate. Plans to build Fukushima I reactors numbered 7 and 8 were abandoned but plans for 11 others in Japan remain subject to satisfactory outcomes from the OECD 'stress tests'. If implemented, such plans would make nuclear electricity generation almost 50% of the Japanese total, but public opinion polls have shown that 70% of respondents favour the phasing out of the industry.

It is already apparent to those appraising the Japanese industry that, while some tolerance of seismic events may have been achieved in the construction of reactors, it was insufficient, and that the design basis for protection from tsunami was never adequate. Ten metres high tsunami waves were expected to occur every 30 years.

Prime Minister Naoto Kan must have been advised after the tsunami of some of the worst outcomes of loss of control of four reactors and spent fuel ponds. The owners of the Fukushima I site early in the crisis planned to abandon it. Kan speculated that in that event millions of people would leave greater Tokyo and that the Japanese economy would collapse. We are told, but we don't yet know by whom, that he learned that further explosive releases of radionuclides to the environment could make large tracts of land uninhabitable – a prospect that he described as intolerable. No doubt he proposed massive curtailment of the industry, and his resignation as Prime Minister may have been for lack of cabinet support. Like Mikhail Gorbachev, he was shocked to learn the hazards of the industry after a disaster. Gorbachev bitterly reproached Soviet nuclear industry leaders for their claims of a safe industry and their failure to describe even to a national leader the nature of failure.

The reactions of other communities to the disaster

In Germany, plans to phase out the nuclear industry have been revived, and in Italy 95% of those polled favoured a non-nuclear energy sector. Switzerland and other countries have, or are expected to follow suit.

In Britain, there are reports of active efforts by nuclear industry lobbyists attempting to influence the reporting of nuclear matters. ('We must work together on this and have a very strong co-ordinated message'; the e-mail author's name redacted.[1]) From the BBC's lack of scepticism in repeating incredible Japanese accounts of 'safe' discharges to the environment, they seem to have succeeded.

The British government, already committed to its justification of nuclear 'new build', must have been grateful for Interim Conclusion No 1 endorsed in Dr Weightman's Final Report that

> *In considering the direct causes of the Fukushima accident we see no reason for curtailing the operation of nuclear power plants or other nuclear facilities in the UK.*

The Secretary of State lost no time in informing Parliament that the report reassures us that

> *... new nuclear can be part of a low carbon energy mix in the UK.*

Dr Weightman, in his May 2011 Interim Report, invited comment on his interim conclusions and recommendations. It was apparent to me that many statements about 'safe' exposure to radiation were being repeated by British media without challenge, and I responded by suggesting that Dr Weightman could quote the majority view of the Committee Examining Radiation Risk of Internal Emitters (CERRIE 2004) and other authorities that there is no safe level of radiation, and that the effects of low exposure are rarely immediate. This was only part of a concern that hazards were being understated. I was relieved to find in the final report a mention of 'genetic effects to progeny', which has had little mention since the management of Sellafield advised its workers to think twice about having children.

The Final Report is detailed and reassuring in its comparison of nuclear regulation in the UK with that of Japan in the design of reactors and in site licence conditions, for example, all showing that British standards are higher. But there is some lack of congruence between the conclusion quoted above and the many detailed recommendations on British practice on such topics as flood protection, the provision of back-up water supplies for spent fuel stores, and planning for emergencies. My interest grew when I read a discussion of proposed changes in Japanese government agencies to prevent conflicts of interest by separating regulation of the industry from promotion of the industry. There was no mention of a similar need in British structures, and this is discussed below.

Towards the end of Weightman's Final Report, in 'Annex L' with the title 'Severe Accident Progression', there are several pages of speculation about the extent of reactor core meltdown, sometimes referred to as 'core relocation'. The limitations of this section, by the sheer lack of information about the status of the three reactors, is freely admitted, but it goes some way to deal with my concern that if heads of state do not know what the worst outcomes of a nuclear disaster can be, there is a need to describe them soon, fully and to the public. Part of the explanation could be that the industry itself does not know, or lacks agreement on, the worst outcomes. Below is an extract from page 270 of the Final Report followed by an extract from recommendation 25, which deals with the same topic.

> *It should be noted that computer code models for vessel failure cannot be considered to be well validated, due to the lack of an experience base against which to benchmark the codes. It should also be noted that MAAP* [Modular Accident Analysis Program] *and MELCOR [Methods for Estimation of Leakages and Consequences of Releases] do not have models for some phenomena discussed above, such as steam explosions.*

From Interim Report 25 (confirmed in the Final Report)

> *the industry needs to ... ensure it has the capability to analyse severe accidents to properly inform and support on-site severe accident management actions and off-site emergency planning. Further research and modelling development may be required;*

Steam explosions were discussed in other sections of the report, as were hydrogen explosions. A search for 'fission explosion' in the 300 page Final Report and its relevant references produced nothing.

The Final Report's end-note reference No 2 with the title 'Report of the Japanese Government to the IAEA Ministerial Conference on Nuclear Safety June 2011' is available via the URL. It is one of the sources used by Dr Weightman for his Final Report. In the Japanese government report shortcomings in design basis standards for reactors are freely acknowledged, as are the failures of the regulatory agencies in not calling for higher standards.

Chernobyl and Fukushima: explosions compared

Zhores Medvedev, in a preface to the 2011 edition of *The Legacy of Chernobyl*^{2}, describes the Chernobyl explosion as a nuclear explosion.

> *With this design, during the first seconds after the 'panic button' was pressed, 170 rods started to move down at once, slowed by having to displace water, not absorbing neutrons, but instead producing a reactivity in the lower part of the*

reactor core, resulting in the explosion due to the increase in criticality and reactivity. The operators did not know about this possibility and it was the first time in Toptunov's short life that he had used the emergency button. The conclusion of the paper in Atomnaja Energia *was that the dramatic increase in reactivity (nearly 100 fold) was a direct result of the design error.*

Medvedev records that the control rods were subsequently redesigned in all RBMK 1000 reactors. At page 33 there is a discussion of the possible extent of meltdown and the desperate measures taken to control it. There is also what is missing from the UK Tolerability of Risk document[3] – an account of the possible consequences of a core meltdown.

The first two explosions were great disasters, but the continuing emissions for many days of fresh radionuclides represented an even greater danger to the population and to the environment. A meltdown of the core would lead to unimaginable damage. If it could not be prevented half of the Ukraine and Byelorussia would have to be evacuated. The land would be contaminated for many years. The Dnieper and other rivers in the area would be affected for many decades. The three other reactors on the Chernobyl site (which were still working and producing electricity) would be destroyed, causing untold further damage. There were about 3000kg of accumulated plutonium and 700,000kg of uranium in the fuel elements of the four reactors in Chernobyl and vast amounts of other radionuclides. Everyone who was involved in the emergency operations around Chernobyl recognised the gravity of the situation but no one knew how to prevent the catastrophe of meltdown.

George Monbiot in a *Guardian* article[4] seemed to have taken some comfort from the possibility that three reactors at Fukushima Dai-ichi suffered some meltdown without the horrendous consequences described above. He declared 'The unpalatable truth is that the anti-nuclear lobby has misled us all'. Some are impressed, the nuclear lobby in particular, by Monbiot's speculation, but if explosive nuclear fission is possible in reactors and spent fuel stores which are out of control, there is little assurance to be gained from the experience that it occurs in one instance out of four and then with less than its full potential. Such consequences are intolerable by any measure, and low probability does not change that. The carefully worded recommendation IR25 of the Chief Inspector's Final Report supports my observation that more information is needed.

Before we leave *The Legacy of Chernobyl,* here is an extract from Dr Medvedev's conclusion of chapter 1 with relevance for the lessons of Fukushima.

It is obvious, however, from an analysis of the safety tests and other major features of the Chernobyl RBMK reactor that the main liability of the system was (and still is) the absence of protection from station blackout – in other words the loss of on-site electric power.

One does not need an earthquake or a tsunami to bring that about. Grid failure in the UK could occur by several causes including terrorism and other hostile action. It is significant that terrorism is not mentioned in the Final Report and that Dr Weightman does not discuss nuclear policy issues. He states

As with the Interim Report, this Final Report does not examine nuclear policy issues. These are rightly matters for others and outside my organisation's competence and role.

This could be a consequence of misdirection of the Nuclear Installations Inspectorate by the Chair of the Health and Safety Executive (HSE), Judith Hackett, when she spoke of a duty to reassure the public[5] but has not explained where such a duty can be found in law or elsewhere.

Sections 2 and 3 of the Health and Safety at Work etc. Act require an employer to provide a safe system of work so far as is reasonably practicable and the nuclear inspectorate is appointed to enforce those requirements. An employer with a choice of methods of generating electricity and who does not wish to produce plutonium is guided by the Act towards the safer methods and, likewise, should be guided by the HSE. When presented with these arguments, Judith Hackett rejected them. By inventing a duty to reassure, and ignoring the central requirements of the Act, these rank as the worst misdirections of the inspectorate since her predecessor declared deregulation to be HSE's priority. These are instances of the independence of the regulator being compromised by government and they require the remedy (structured independence) advised and already accepted by the Japanese government.[6] The restructuring of the UK Nuclear Inspectorate, when no defect in their performance has been described, appears to be taking place ahead of legislation.[7]

In this context, it is important to note that action by terrorists is excluded from the Stress Tests proposed by the European Council and that the exclusion was at the insistence of the British Government.[8]

Other experts who postulate nuclear explosions in reactors

A Dr Webb, formerly of the United States Navy, who had worked on nuclear submarine reactors, preceded me in questioning J D Rimington on his evidence on the 'Tolerability of Risk from Nuclear Reactors' at the Hinkley Point 'C' public inquiry[9]. (Mr Rimington's document once having been mistakenly quoted as the 'Risk of Tolerability Document'.) Dr Webb was very unpleasantly aggressive in his questioning, and I remember that I wished not to be associated with it. But the record shows that Dr Webb's assertion that nuclear explosion was possible was not disputed by Mr Rimington, nor by two senior HM Inspectors of Nuclear Installations. They treated the suggestion as incredible, but stopped short of saying that it was impossible. The NI inspectors undertook to read Dr Webb's texts, but Mr Rimington refused Dr Webb's request that he be informed of their conclusions. Dr Webb did establish that Mr Rimington was neither a scientist nor an engineer. The Tolerability of Risk document offered probabilities of a 'loss of coolant accident' but did not make clear what the consequences of such an accident could be.

The late Professor Jack Harris, FRS, FEng, was a nuclear metallurgist involved in the design of British gas cooled reactors, and a vice chairman of the British Pugwash Group dedicated to the elimination of nuclear weapons. As a supporter of Pugwash, and as a colleague in the Institute of Materials, Minerals and Mining, I was drawn to him because of his support for Ross Hesketh's position on the possibility of a nuclear explosions in nuclear reactors.[10] He and I exchanged several letters and e-mails and he confirmed that opinion, which he had already published, some years later, shortly before he died.

I worked with UK nuclear installations inspectors as HM District Inspector of Mines and Quarries when the Health and Safety Executive created a working party on management in high-risk industries, in response to the recommendation by the late Sir Frank Layfield at the Sizewell Public Inquiry. (My work included the enforcement of the Ionising Radiation Regulations.) The working party prepared a report on human error which, it became apparent, was not quite to the liking of the then Director General, Mr J D Rimington. My contribution on human error was to say that experienced managers expect it.

In retirement, I attended a seminar series at the London School of Economics and noted the response of the then NII Chief Inspector, Dr Sam Harbison, to the concern of the Association of Nuclear Free Local Authorities that a failure at a British nuclear installation could render large parts of Britain uninhabitable. He did not dispute the advice given to NFLAs. He described such failure as 'a low probability event with a high outcome'.

Until recently, I had little information about how many other Fellows of the Royal Society shared Jack Harris's view on the possibility of nuclear explosion in a reactor. Even the President of the Royal Society replied that he had little information. Advocates of nuclear electricity generation had given me the impression that such explosion was impossible, but HSE's responses to Dr Webb's questions in the verbatim record of the proceedings of Day 59 of the Hinkley Point 'C' public inquiry at Cannington persuaded me that I was wrong about that.

Jack Harris was not the only Fellow of the Royal Society persuaded of the possibility of explosive fission in reactor fuel. It was the then secretary of the Association of Nuclear Free Local Authorities who provided me with the source of the statement by Sir John Hill when Chairman of the Atomic Energy Authority in the UK. He wrote in the house journal of the Authority in 1992

When the Americans chose graphite moderated water cooled piles for plutonium production they recognised that a failure of the water supply or control system could result in prompt criticality and a nuclear explosion such as happened 40 years later at Chernobyl.[11]

Wikipedia and recent debate

Wikipedia articles, which can be edited by any person, now enjoy some protection from vandalism and are a useful source of information. Topics that are subject to peer review in science and technology journals are better sources, but often require passwords and costly subscriptions for lay readers. The Wikipedia article on Chernobyl contains the following

A second, more powerful explosion occurred about two or three seconds after the first; evidence indicates that the second explosion resulted from a nuclear excursion.[33]

The expression 'nuclear excursion' appears in blue with underlining indicating that it is the subject of a separate article. The end note reference suffix (33) leads to

Pakhomov, Sergey A; Dubasov, Yuri V; (16 December 2009). 'Estimation of Explosion Energy Yield at Chernobyl NPP Accident'. Pure and Applied Geophysics (Springerlink.com) 167 (4-5)

with a digital object identifier reference number, which leads to the complete paper with open access. The following is from the abstract

Comparison of estimated results with the experimental data showed the value of the instant specific energy release in the Chernobyl NPP accident to be $2 \cdot 10^5 – 2 \cdot 10^6$ J/Wt or $6 \cdot 10^{14} – 6 \cdot 10^{15}$ J (100–1,000 kt). This result is matched up to a total reactor power of 3,200 MWt. However this estimate is not comparable with the actual explosion scale estimated as 10t TNT. This suggests

> *a local character of the instant nuclear energy release and makes it possible to estimate the mass of fuel involved in this explosion process to be from 0.01 to 0.1% of total quantity.*

The separate article has the title 'Criticality Accident'. It states that

> *Although dangerous, typical criticality accidents cannot reproduce the design conditions of a* fission bomb, *so* nuclear explosions *do not occur.*

This represents the views of those contradicting Medvedev, Pakhomov and Dubasov, and the incompatibility of this article with the current Wikipedia article on Chernobyl is pointed out in the 'talk' page of the separate article and remains unresolved. It cannot be resolved if, by definition of 'Criticality Accident', bomb-like criticality is excluded. Resolution requires more than the begging of the question. The bomb makers could help with this, if they were not sworn to secrecy. In the meantime, laypersons like me see the debate proceeding 0.1kt of TNT equivalent at a time. We cannot be content that perhaps only 0.1% of a reactor's energy resource is acknowledged to be capable of atomic bomb-like explosion.

It is significant that the conclusion of the current Chernobyl article, that a nuclear explosion occurred approximately equivalent to 10 tonne of TNT, is not contested by any recent edit. A perusal of earlier versions shows that there has been no contest for at least the last six months. Edits of other detail have been made at the rate of up to ten per month.

What the judges said

Mr Justice Sullivan, in the High Court on 15 February 2007, ruled that the UK government's second consultation on energy policy was 'seriously flawed' and thus 'unlawful'. There had been no consultation at all, he said, because the government had provided information 'wholly insufficient for the public to make an intelligent response'.[12] In fact, the government had also blacked out the economic data in papers obtained by the provisions of the Freedom of Information Act.

The conclusion in HM Chief Inspector's Final Report on Fukushima that there is no reason to desist from building more nuclear power stations is barely supported by the findings and recommendations of the report, some of which convey well enough the need for more information on the worst outcomes of nuclear disasters.

People who remember that employees at Sellafield were once advised by their employer to think twice about having children could be a minority by now, as are those who recall that we once had public inquiries into all the controversial and unresolved societal aspects of nuclear industry matters. Loss of habitation and infrastructure, loss of habitable land, loss of agricultural land, of clean water supplies, of animals and plants and their genetic integrity are hardly mentioned in the Tolerability Document. Similarly, more explicit and quantified accounts of harm are required of the Environment Agency when appraising new processes and the releases to the environment that will occur. Changes initiated by the Blair Government have made public inquiries unlikely as part of the 'fast track' process of 'New Build'.

Misleading information

'Safety is no longer an issue' was the statement volunteered by David Cameron as leader of the opposition to Tony Blair as Prime Minister in a 2006 debate on nuclear energy policy. No doubt Tony Blair was grateful at the time, but both can now regret that they made themselves hostages to the misfortune of Fukushima Dai-ichi. They should have known the vulnerability of reactors and spent fuel stores to loss of power supplies.

The publication of several photographs of an interim nuclear waste storage facility at shallow depth, wrongly described as an underground disposal facility, is an uncorrected example of the last government's attempt to support its assertion that 'solutions' exist for waste management problems.[13] My review of *Geological Repository Systems for the Safe Disposal of Spent Nuclear Fuel and Radioactive Waste* was published in *Materials World,* the journal of the Institute of Materials, Minerals and Mining.[14] I found that the papers by the 23 author teams and the two editors of this 750 page book did not concur that we have solutions for the management of highly active spent fuel waste. They made clear that nowhere in the world does a functioning geological depository for such waste exist, and the authors express caution rather than consensus on the likely availability of a depository for safe long-term containment.[15]

'The AP1000 is the safest and most economical nuclear power plant available in the worldwide commercial marketplace.'[16] A separate invalid claim made by the Westinghouse Company that the AP1000 reactor exists as 'a proven design already built elsewhere in the world' is dealt with in my booklet *Nuclear New Build – a Review of the Issues* at page 43.[17] A third invalid claim that the AP1000 reactor exists, this time by the Nuclear Industry Association endorsed by government, provides detail:

> *The Application sets out in detail the technical features of the AP1000 ... that the nuclear systems are located in the shield building/containment vessel and in the auxiliary building. These buildings are robust and shielded where necessary to ensure all radioactive substances are always secure.*[18]

The AP1000 reactor does not exist, and its prototype AP600 was never built. It strains belief that such false statements were sufficient to found Tony Blair's policy of building ten non-existent reactors as a 'fast track' programme to mitigate global warming. In the last five years the company has been unable to submit an acceptable design to the Nuclear Inspectorate and, since Fukushima, has retreated from the Generic Design Assessment (GDA) process for lack of funds. Mendacity as a characteristic of the nuclear industry appears to have extended itself into government. The fact that it does not insure itself for more than 1% of the potential claims carries it own message about nuclear safety.[19]

The argument that global warming and climate change require the pursuit of low carbon electricity is sound. That nuclear electricity is the only way is contradicted by the 2002 Energy Review, which found renewable energy available and sufficient for reasonable economic growth. We have lost a whole decade in developing those resources with the vigour that was needed, and capital investment now in 'new build' can only delay the benefit of benign renewable energy. Tony Blair's reasons for rejecting the 2002 Energy Review, when the industry was far from ready for expansion, remain to be explained.

Public Inquiry

The gist of this article is that the principal recommendation of the Dr Weightman's Final Report on Fukushima, that he sees no reason to curtail the operation of nuclear power plants and other nuclear facilities in the UK, is not supported by the findings and recommendations of the report, which acknowledge a failure to describe the potential worst effects of loss of control of reactors and other facilities.

The Secretary of State was quick to use the report as validation of his Statutory Justification decision of new build – that the benefits will outweigh the detriments. Parliament may not agree, and it would be prudent of the Secretary of State (who is personally opposed to new build) and the government, if they wish to enjoy the confidence of the public, to initiate a public inquiry to re-assess the detriments of nuclear power generation. In the absence of a public inquiry, Greenpeace's application for judicial review of the Justification decision will surely find many more supporters.

References

1. 'Emails released under the Freedom of Information Act reveal the level of coordination between government departments and the nuclear industry during the Fukushima crisis' 'We must work together on this and have a very strong co-ordinated message' (The e-mail author's name redacted) Guardian 30 6 11
2. *The Legacy of Chernobyl* Zhores Medvedev Page iv Spokesman Books.com 2011 £19.95
3. *The Tolerability of Risk from Nuclear Power Stations* Proof of Evidence by J D Rimington, Director General of the Health and Safety Executive, to the Hinkley Point 'C' Public Inquiry. Annex 1 to Inquiry Document Reference HSE 1.
4. 'The unpalatable truth is that the anti-nuclear lobby has misled us all' *The Guardian 5 9 11*
5. Statement by Judith Hackett when opening the Nuclear Division conference on 17 July 2009. She said 'It is essential that we build and maintain public confidence in a safe nuclear future – and an independent regulator is an essential element of that.'
6. *The Guardian* 16 8 11 Editorial 'After Fukushima – Nuclear Dirty Tricks' (in which Angela Merkel is described as one of the few political leaders who is also a scientist)
7. Gifford, C *Nuclear New Build – a Review of the Issues* p36 Spokesman Books Dec 2010 £6-00 The Bertrand Russell Peace Foundation Russell House Bulwell Lane Nottingham NG6 OBT 0115 9708318.
8. Leigh Phillips writing in *The Guardian* 26 5 11
9. Transcript of Proceedings of the Inquiry on Day 59, 31 January 1989, pp96-107 on the cross-examination of Mr J D Rimington, Director General of the Health and Safety Executive. Social Studies Library The University of Wales, Colum Road, Cardiff UK.
10. *Ross Hesketh* Jack Harris in *Materials World* a journal of the Institute of Materials, Minerals and Mining, June 2004 London
11. *Atom* No 421 March 1992, page 7
12. The High Court 15 February 2007 in Judicial Review on the application of Greenpeace and others of the government's second consultation on energy policy.
13. In the government publication 'Managing Radioactive Waste Safely – a Framework for Implementing Geological Disposal' June 2008 three photographs illustrate 'an underground disposal facility'. The photographs are of the Swedish Forsmark facility which the developers SKB describe as 'an interim storage facility at shallow depth.' Was this three innocent mistakes by government or another dodgy dossier?
14. February 2011
15. My paper *Geological Disposal of Nuclear Waste* was presented at a meeting of

the Wales Branch of the Institute of Materials, Minerals and Mining on 16 September 2008 at the Cardiff University Department of Earth Sciences. 28pp 58 end notes. It is posted on the website of the South Wales Institute of Engineers Educational Trust http://www.swieet2007.org.uk.
16 From the Westinghouse website.
17 Gifford, C *Nuclear New Build – a Review of the Issues* Op cit.
18 *The Justification of Practices Involving Ionising Radiations Vol 2*, p23, DECC Nov 2009. The Application mentioned is that by the Nuclear Industry Association to justify new nuclear power stations.
19 My response to the DECC Consultation on the Revision of the Paris and Brussels Conventions on Nuclear Third Party Liability 20 April 2011.

UK Energy Policy?
The nuclear dimension

Ian Fairlie

This article was published in *The Spokesman 124: Problems of NATO* (2014).

Dr Fairlie is an independent consultant on radioactivity in the environment, based in London.

The Government's proposed deal at Hinkley C with Electricité de France would hand effective control of a major part of the UK's nuclear electricity to the French and Chinese Governments, double the price of electricity, and result in thousands more old people dying from fuel poverty.

In recent months, the UK's energy policies have rarely been out of the headlines. We have seen dashes for gas; an all-out push for nuclear; yo-yoing on renewables; and much hot air on energy efficiency subsidies. In December 2013, the Prime Minister asked the 'Big Six' energy companies to lower energy price increases, following the Labour Party's promise to freeze prices if elected in 2015. Add the Government's support for unpopular fracking, together with the widely criticised proposed deal on nuclear electricity at Hinkley C, and one begins to wonder what is going on. In Europe, energy analysts apparently shake their heads in disbelief at the disjointed series of events occurring in the energy sector in the UK.

This is alarming, as coherent energy policies are vital to address several important issues, including security of supply, global warming, and fuel poverty. Just to take the last point: 31,000 elderly people die every winter from fuel poverty, according to the Office for National Statistics, which amounts to eight older people every hour in winter, from NHS figures.

Several factors contribute to the muddle. One is the existence of strong differences of opinion within the Conservative/Liberal Democrat Coalition on most aspects of energy policy. Another is that the Government is repeatedly outmanoeuvred by the Big Six energy companies, though the Department of Energy & Climate Change (DECC) and the Treasury refuse to admit this. It doesn't help that the energy companies reportedly have numerous advisers inside DECC steering energy policies their way. A further one is that the Government is ideologically bound in its resistance to more regulation when that's what is needed. This is similar to the Government's adherence to market dogma, despite clear evidence that the market has often failed in this policy area – namely, carbon pricing, energy efficiency projects, contracts for difference, even global warming itself, according to the former Treasury advisor, Lord Stern. What emerges here is a picture of a dysfunctional and inconsistent framework for implementing national energy aims.

One energy policy area stands out as particularly incoherent: nuclear power. To say that the Government appears besotted with nuclear is an understatement: it's more like an obsession. Consider the following. The Government wants to give £10 billion to the French Treasury (via its ownership of 84% of Electricité de France (EdF)) to pay for 2/3rds of the cost of building the proposed Hinkley C plant in Somerset. It wants to give a major say to the Chinese and French governments in building and running the plant. And it wants to promise EdF to double the current price of nuclear electricity and guarantee this for 35 years – worth another estimated £40 billion to the French Treasury. Recently, the British Government announced it also wanted to offer similar massive subsidies to Toshiba (which owns US Westinghouse) for a new nuclear power station in north Wales. These policies, if implemented, would have major adverse effects on UK fuel poverty and winter hypothermia deaths, at the least.

Indignant public responses

Perhaps 'incoherent' is too polite a word. Certainly, the public's response to DECC's proposed deal with Electricité de France on nuclear electricity prices, announced in October 2013, has been highly indignant. Here are some of the more colourful raspberries:

the Energy Secretary had let EdF *'take the British Government for a ride'* over the *'ludicrously high'* subsidy deal to fund the proposed £16 billion Hinkley nuclear plant. Lord Lawson, *former Chancellor (8 November 2013)*

'Flabbergasted ... we are frankly staggered ... Hinkley will be the most expensive power station in the world.' Peter Atherton, *Liberum Capital (30 October 2013)*

'We could be staring at a truly astronomical cost by the end of the contract.' 'The government surely can't be that dumb,' comments one

City analyst. *'One assumes not.'* Nils Pratley, Guardian finance writer, (18 October 2013)

'Hinkley – a lousy template for nuclear Britain ... it's hard not to have misgivings over the costs and strategic logic of this deal ... by 2023, consumers will be paying £720m a year above the market price ...' Alistair Osborne, Daily Telegraph finance writer (21 October 2013)

'... a huge public contribution towards yesterday's energy thinking.' Alan Simpson, former Labour MP (23 October 2013)

And there are plenty more in the same vein.

Nuclear fetish – why?

The British Government's nuclear fetish is hard to understand, given the prohibitively high costs of nuclear power. Nuclear construction costs have always been high but, in recent years, they have increased substantially: the anticipated cost of Hinkley C is now £16 billion, which is 1.5 times the cost of the 2012 UK Olympic Games. This is for one nuclear station which would supply less than 4% of the UK's electricity if it were ever built and operated.

Currently, two nuclear stations are being built in Europe: both are wildly over budget and years, approaching decades, behind schedule. These are European pressurised reactors (EPR) – the same type EdF wants to build at Hinkley. Major legal, financial and technical questions hang over both European projects: they may well never be finished. The plant under construction in Finland at Olkiluoto is in severe difficulties, and it is thought that Areva, the constructors, may even have pulled out of the project. The other is in France.

Meanwhile, the costs of renewable energy sources such as wind power and photo-voltaics continue to plummet. As a result, nuclear projects across the world are increasingly being abandoned. For example, Germany, Belgium, Switzerland, Italy and Japan, as well as large multinationals such as E.ON, RWE and Siemens have abandoned nuclear in order to pursue renewable energy policies. As pointed out by *The Economist* ('Britain runs towards nuclear energy as other countries flee', 26 October 2013), the UK is alone in the European Union in having advanced plans for more nuclear power, apart from Finland.

On 21 February 2014, *The Spectator* followed up by asking 'Why has Britain signed up for the world's most expensive power station?' It stated that MPs owed it to the taxpayer to throw out the proposed Hinkley deal.

Justification for nuclear: climate change?

The British Government's justification for new nuclear is that its low CO_2

emissions address concerns over climate change. There is little doubt that global warming is a real and serious threat, but nuclear is a poor answer. Those who defend nuclear, including columnists such as George Monbiot and former environmentalists such as Mark Lynas, appear to overlook that uranium mining, uranium milling, uranium enrichment, nuclear fuel fabrication, and radioactive waste treatment all have heavy carbon footprints.

But it's more than that: the crucial factor is that nuclear has limited potential to reduce UK's CO_2 emissions. In 2006, the Government's former Sustainable Development Commission estimated that a 10 GW fleet of new nuclear power stations would address 4% to 8% of the UK's carbon emissions depending on assumptions. This means Hinkley C alone would address ~0.5% to ~1%.

In fact, of the many options available (wind, wave, solar thermal, photovoltaic, biofuels, hydro, etc) nuclear is arguably the least effective way to reduce CO_2 emissions. Amory Lovins, the eminent US energy guru, has calculated that, in terms of $ per tonne of C saved, nuclear is the worst possible way to reduce CO_2 emissions: efficiency and the renewables are much better methods. Moreover, the cost gap between nuclear and the renewables is widening daily.

Apart from its past history of distortions, cover-ups and secrecy, the nuclear industry suffers from other disadvantages. The ongoing crisis at Fukushima in Japan following the quadruple explosion, triple meltdown nuclear accident in March 2011 is not reassuring. Neither is nuclear sustainable, following Cumbria County Council's decision in February 2013 to oppose DECC's plans for dumping nuclear waste in Cumbria. And there is always the spectre of nuclear proliferation worldwide.

What we are missing?

In much of Europe, and even nowadays in the United States and the developing world, 100% renewable energy goals are becoming the norm. Tragically, the British Government's nuclear plans mean we are missing out on the following electricity potentials:

[TWh= terawatt-hour (10^{12} watt-hours) the unit for electricity generated/used]

155 TWh/year generated by offshore wind;

40 TWh/year by implementing a comprehensive domestic energy efficiency programme by 2030;

100 TWh/year saved through other efficiency measures;

22-140 TWh/yr from solar PV on domestic roofs;

30 TWh/yr from solar PV on industrial and commercial roofs;
140-190 TWh/yr from solar farms – just using land currently used for growing biofuels.

This totals ~500 TWh/yr which is greater than the UK's current consumption of ~330 TWh/yr.

Recent changes

Recent months have seen increased questioning of the Government's mania for nuclear, as shown by the colourful comments in response to the proposed deal at Hinkley. On 28 November 2013, former CEO at BP and current Government adviser, Lord Browne, stated that nuclear power was 'very, very expensive indeed'. He reflected the views of some bankers such as the President of the World Bank who, in response to questions about the bank's energy lending policies, stated 'we don't do nuclear energy'. Even *The Times,* in December 2013, published a letter critical of the government's absurd nuclear plans from a dozen academics.

But it's hard to be optimistic about the prospects for immediate change. One problem is the disinformation peddled by some newspapers and media about UK renewables, and about Germany's decision to embrace renewables and exit nuclear – a policy that we would do well to emulate. How many know that Germany now has more than 450,000 jobs in the renewables industries compared with about 30,000 here?

Finally, it's dispiriting that the Labour Party is just as attached to nuclear as the Con-Dem coalition. Labour's acceptance of the proposed Electricité de France deal at Hinkley means cross-party agreement exists which is difficult to dislodge even if it's absurd. In other words, more old age pensioners will die from fuel poverty in future if the deal is implemented and electricity prices were increased. It's high time Labour's pro-nuclear policy was re-examined: we should recall that, prior to 2003, the Labour Party and the TUC were formally anti-nuclear. What happened then? Tony Blair forced through a pro-nuclear policy.

Submerged politics of UK nuclear power

Is Trident renewal influencing UK energy policy?

Phil Johnstone
Andy Stirling

This article was published in *The Spokesman 132: Another Europe is Possible* (2016).

Philip Johnstone is Research Fellow, Science Policy Research Unit, University of Sussex. Andy Stirling is Professor of Science & Technology Policy, SPRU, and co-director of the ESRC STEPS Centre, University of Sussex.

With Parliament now getting ready to vote on the 'main gate' decision on renewal of the Trident programme, 2016 is set to be a decisive year for the future of UK nuclear weapons capabilities. Political opposition has grown in Parliament, with both the Scottish National Party (SNP) and Labour leaderships now opposed to Trident renewal. At a lifetime cost variously estimated between £31 Billion[1] and over £100 billion[2], the political and economic stakes are very high. Debate is becoming increasingly heated over the practicalities, costs, ethical and strategic implications. Many of these arguments are covered extensively elsewhere, and are not repeated here.[2-6]

Instead, this article looks at another possible implication of Trident renewal which has remained almost completely 'under the radar' of contemporary policy and academic debate. This concerns the recent history of the UK civil nuclear power industry, which also involves remarkably similar stories of delays, cost overruns, questions of necessity and performance, and critical comparisons with strategies in other countries and arguments for superior alternatives.[7]

The intensity of UK commitments to civil nuclear power is also looking increasingly anomalous on the world stage. The contrast with Germany is especially striking, with the UK hosting a massively less successful nuclear engineering and power industry and enjoying a renewable resource that is the envy of Europe.[8] Yet it is Germany (with a track record of prescience in past industrial policy decisions), that is undertaking a complete nuclear phase-out by 2022, whilst the UK Government doggedly pursues a 'nuclear renaissance'. In a current academic research project now nearing completion[9], we are systematically exploring possible reasons for the UK's internationally-anomalous commitment to nuclear energy. And this is where there emerges a seeming connection to Trident.

Of course, concerns over climate change and energy security certainly play a part in UK interest in nuclear power. But they do not explain why the UK should be so unusually intense in its nuclear enthusiasm. As in Germany, such reasons might speak even more strongly for alternative policies. In our research, we have (like others) examined in great detail, issues of energy economics, industrial policy, available resources, security of supply, political lobbies, the history of energy institutions, technological lock-in, and different aspects and qualities of democratic decision making.[8] Although the issues are highly complex and any full explanation must be multi-causal, it is difficult to avoid recognising that there emerges a further factor – one which is all the more important to address, because it has hitherto escaped virtually any attention whatsoever.

In short, these neglected questions concern the extent to which UK policy commitments to nuclear power reflect a deeper perceived imperative to maintain national capabilities to design, build, operate, staff, regulate and decommission nuclear propelled submarines. Without nuclear propulsion, submarines would not, in current military opinion, display the requisite endurance, stealth, speed and robustness to serve as credible platforms or guardians of strategic nuclear capabilities.[10] In influential quarters, capabilities to maintain naval nuclear propulsion is thus seen to constitute a serious bottleneck in the sustaining of crucial wider strategic military capabilities. And these are in turn of crucial importance to a particular UK identity as an 'outsized power' that 'punches above its weight'.[11]

The challenge is that nuclear submarines are among the most complex and demanding of human artefacts. In a time of serous decline in UK manufacturing capacities, maintaining this capability places especially serious demands.[12,13] The security sensitivities preclude much of the kind of national outsourcing that is so routine in other industries. So, the ability of the UK to maintain a cherished elite identity on the world stage, rest on its ability to find as many alternative ways as possible to secure the national reservoirs of highly specialist expertise, education, training, skills, production and regulation necessary to sustain nuclear submarines. In order to achieve this, however, it is not essential that the UK take a lead in building civilian nuclear power reactors. All that is required is that crucial parts of the submarine industry secure key places in civilian nuclear power supply chains.

If this is a factor in the peculiar intensity of UK government commitments to civil nuclear power, then what is most remarkable is that it remains entirely unacknowledged in any policy literature that we are aware of concerning the formal rationales for a UK 'nuclear renaissance'. It would perhaps be in the nature of such a sensitive imperative, that the Government might be expected to be discrete about it. Yet we believe we have found strong circumstantial evidence, that this actually forms a crucial general pressure that has operated decisively at important critical junctures in UK nuclear policy making. It is this evidence that the rest of this article examines.

A military nuclear connection, in this day and age?

It is for good reason that something of a taboo has arisen over the years around emphasising any kind of linkage between civilian and military-related nuclear issues. The topic is the object of much misleading casual comment. Albeit not perfect, strict safeguards have been in place for decades to prevent cross-overs in usage of fissile materials and ensure that civilian nuclear power does not compound risks of nuclear weapons proliferation.[14] Dedicated institutions like the International Atomic Energy Agency (IAEA) and Euratom work strongly to ensure the separation of civilian and military related nuclear matters and uphold the Nuclear Non-Proliferation Treaty.

Perhaps even more significantly, the reduction in strategic arsenals following the end of the Cold War means that key nuclear weapons materials such as plutonium are actually in surplus on the military side.[15] The situation is arguably a little more complex and obscure with regard to other specialist materials such as tritium[16], but with many other possibilities in play, this also seems largely irrelevant to any pressure to maintain a large indigenous civil nuclear power industry. So, although the history of nuclear power in the UK (as elsewhere) is inextricably tied to ambitions around nuclear weapons[17,18] – and the connection remains relevant around horizontal proliferation – it is not credible to argue that nuclear weapons materials production might currently constitute a significant driver of UK civil nuclear policies.

But this is not a story concerning fissile materials. Nor is it about the design or manufacture of vital missile or warhead components, many of which are supplied by the United States.[3,4] Indeed the issue here is not about nuclear weapons at all, but about the ability to construct and operate the submarine platforms on which their effective strategic performance is seen to depend. And – although also linked in many ways to US designs and supply chains – it is an ability to maintain minimal independent national capabilities to build and operate these nuclear-propelled Trident submarines (and associated attack boats) that remains the focus of an intense and rather anxious national debate on the military side.

Here, a long series of government reports, consultancy studies, select committee inquiries, lobby documents and dedicated new institutions all indicate, very strongly, the weight of priority attached to maintaining a threatened national capability. All that is missing, is any clear policy acknowledgement that it is this perceived imperative that is exerting an influence on the strength of commitment to maintaining a UK supply chain sustained by a civilian nuclear power programme. But we think we have found some illuminating tell-tale signs of such links.

The intensity of the UK's commitment to civilian nuclear power is puzzling

Before turning to these, it is important to substantiate quite how distinctive are the current levels of UK Government support for nuclear power. With Energy Minister Amber Rudd recently stating that '*[i]nvesting in nuclear is what this Government is all about for the next twenty years*'[19], the UK is the main governmental advocate on the world stage of a 'nuclear renaissance'. A few countries have larger envisaged programmes in absolute terms[20] (most also, incidentally, operators of nuclear submarines). But these nuclear programmes are much smaller in relative terms when compared with plans in the same countries to exploit low carbon renewable energy options. Globally, investments in renewable electricity generating capacity exceed even that for all fossil fuels put together, leaving nuclear far behind.[21]

Yet UK Government support for a 'nuclear renaissance' remains larger than (and in large part an *alternative* to) efforts to develop its own especially attractive national renewable resources.[22] And what is especially striking here, is how persistent these enthusiasms have remained for a 'nuclear renaissance', despite repeated serious set-backs. The detailed ways in which the UK will deliver on this emphatic commitment are amazingly volatile. Since 2006, a series of radically different designs have each been confidently identified before being abandoned, including designs from US-Japanese, Chinese and French-led consortia – and now, most recently, an as-yet entirely undeveloped US/UK concept for a new small modular reactor.[23]

Nor does past UK history in the nuclear sector offer any encouragement for such an optimistic attitude. Following a series of earlier policy disasters, recent further blows include the withdrawal of multiple earlier prospective reactor constructors[24], massive over-runs in time and cost for similar planned reactors[25], the impossibility of securing private finance[26], the imposing of punishing terms by the current Chinese government financial backers[27], the revelation of a catastrophic defect in a key reactor component[28], and the presently-threatened bankruptcy or withdrawal of the only serious current contender for actually constructing the next UK nuclear plant.[29]

All this has occurred against the backdrop of ample evidence for the ready availability of more cost-effective zero-carbon resources for electricity generation in the UK. Under the same presently-envisaged contracts that are currently evidently viewed as insufficient by the prospective developers of the Hinkley Point C plant, EDF, British electricity consumers will be locked into funding this plant with guaranteed prices over 35 years that are almost three times the current wholesale price of electricity.[30]

The 'strike price' of £92.50/KWh agreed 'behind closed doors' with EDF is significantly higher than the government's own figures for comparable contracts for renewable electricity.[31] And worldwide statistics show unequivocally that nuclear costs continue to rise, whilst global renewable energy costs are falling. National industrial, employment and investment opportunities presented by capital-intensive renewable energy infrastructures are at least equal to those offered by nuclear power. Operational challenges posed by particular renewable technologies such as wind, which are intermittent in their output, are not trivial. But these do not arise until system penetrations that are much greater than presently envisaged scales of development. And they are, anyhow, balanced by a series of countervailing qualities in distributed electricity technologies that are actually seen in countries such as Germany and Denmark as advantages when compared with inflexible centralised 'base load' nuclear power.[32]

Of course, much scope remains for argument on all sides. Energy issues are complex, uncertain and ambiguous. But it is not necessary to be an unqualified critic of nuclear power, to appreciate that it is extraordinarily difficult to reconcile the intensity of UK government commitments to nuclear power with the recent history of experience in this field, neither with established global trends, nor with the manifest cost-effectiveness and availability of low carbon alternatives. Against a backdrop of a stronger national nuclear industry and a weaker national renewable resource,

Germany presents an especially telling contrast. At the very least, it does seem that some other explanation is required for why the UK should remain so internationally distinctive in the intensity of its attachment to a 'nuclear renaissance'.

The 2003 Energy White paper: an exception that proves the rule?

In seeking to understand the causes of this evidently peculiar form of technological lock-in, it is illuminating to consider a brief period when the attachment was briefly broken. After a series of policy catastrophes driven by successive episodes of apparent UK Government credulity in the face of over-optimistic representations of nuclear interests[33-35], the 1997 election saw all political parties, if reluctantly, accepting that nuclear power had become uncompetitive and unattractive compared with alternatives.[36] In the ensuing new enthusiasm for public participation in the early years of Tony Blair's New Labour administration, an unprecedented move occurred when the Cabinet Office initiated an important review of energy policy that was not primarily written by Government civil servants but also included crucial inputs from independent energy experts.[37] Also relevant is that this arrived at its energy focus through a rather convoluted route that began as a review of resources, quickly evolving to include renewable resources, and then expanding to address other energy options more generally. In this way the energy issue was approached 'under the radar', by-passing the 'usual suspects' in established ministries concerned with nuclear strategies.

For whatever reason, the resulting report became the most detailed UK government analysis to date of the imperatives involved in undertaking a transition to low carbon energy systems. Following up on this, the Energy White Paper of 2003 concluded, in an exceptional historic moment, that nuclear power was not an attractive option – and that a shift towards a more decentralised energy system based around renewables and energy efficiency would be preferable.[38]

What followed was one of the most remarkable turnarounds in recent UK policy-making on any issue – offering some of the most compelling circumstantial evidence for the relevance of military submarine capabilities as a driver of civil nuclear policy. In an unprecedented short period after the publication of the 2003 Energy White Paper, Tony Blair announced in 2005 a completely new energy review. Without providing any substantive reason as to its necessity, this further energy review was undertaken by a small group of civil servants in the Cabinet Office. According to one nuclear proponent, Simon Taylor, this involved a select group that most civil servants in the Cabinet Office did not even know existed, working 'in secret' specifically to re-examine the case for nuclear power.[39]

The resulting energy review was thus far shorter than the earlier process, entirely dependent on narrow government specialists, and largely conducted in secret. The consultation for this review was managed by AEA Technology (the former Atomic Energy Authority). Amidst widespread bewilderment and criticism of this superficial process was a finding by the Royal Courts of Justice that the new government consultation was actually 'unlawful' in its bias towards nuclear power.[40] Although by no means opposed to nuclear power, the House of Commons Trade and Industry Select Committee also concluded that the consultation was a 'rubber stamping' exercise to reverse the conclusions of the more rigorous, longer, and independent energy review of 2002-3 and construct an apparent 'need' for new nuclear build.[41] Tellingly, Tony Blair's response to this formidable reaction was that it 'would not affect policy at all'.[42] A second rapid consultation was staged, abandoned by non-governmental organisations as again being flawed[43], and by 2008 a final new nuclear White Paper was released with exactly the same conclusions.[44]

With the rationale for this remarkable turnaround so manifestly determined in such authoritative ways as inadequate, what evidence might there be for alternative explanations? And it is here that our story turns to the apparent links with military submarine capabilities.

Submerged factors influencing UK energy policy?

It was in exactly this 'critical juncture' between 2003 and 2006, that an unprecedented intensification can be seen in policy activities around UK 'submarine nuclear capabilities'. Much of this discussion is internal to the military sector and addresses civil nuclear policy only incidentally. But the overall picture is very clear – it was at precisely the point when civil nuclear power fell out of official favour that anxieties arose in an unprecedented and abrupt fashion that a serious threat had arisen to the ability of the UK to maintain a national capability to build and operate nuclear submarines.

One significant element in this wider series of developments was an extremely energetic and well-targeted initiative by interests associated with the Barrow Shipyard where all UK submarines are constructed – formerly by Vickers and now by BAE Systems. Formed in March 2004, this well-funded group, Keep Our Future Afloat (KOFAC), involved trade unions, local councils, and county councils in concerted efforts to sustain the construction of nuclear powered submarines at the Barrow shipyard.[45]

Targeting politicians and party conferences, producing key reports and submitting evidence to both civilian energy policy reviews and defence reviews, the intense lobbying campaign came to be seen by parliamentarians as 'one of the most effective' that they had ever encountered.[45]

There emerged during this same 'critical juncture' defined by the unprecedented turnaround in civil nuclear energy policy a series of other remarkable indications of the political energy unleashed by concerns over submarine nuclear capabilities. It was in 2005 that the Ministry of Defence funded the RAND Corporation to conduct an in-depth three volume study of the 'nuclear submarine industrial base'.[46-48] Concerns were explicitly discussed over whether the UK would have the key relevant skills to construct nuclear submarines.[49] There ensued a series of Select Committee inquiries into exactly this topic.[10] Evidence was heard from a wide range of interested parties, many of whom explicitly addressed the relevance to the maintaining of UK nuclear submarine capabilities of the parallel sustaining of a healthy civil nuclear industry.

Other reports on exactly this theme were also produced around this time by other bodies including the International Institute for Strategic Studies[50], and the Royal United Services Institute.[51] The latter was led by a senior figure from inside BAE Systems who – among other interesting allusions to linkages between civil and military industries – referred to strategies in other cases under which particular military programmes can be 'masked' in other activities. It was on this basis that the founding moves were made behind major current policy initiatives with missions spanning both military and civilian sectors. The Cogent Programme[52], Key Suppliers Forum[53] and Nuclear Institute[54], for instance, all have explicit responsibilities to protect capabilities relevant to both military and civilian nuclear sectors.

Taken together – and despite the lack of explicit policy acknowledgements – the evidence seems clear. As observed by Oxford Economics in a detailed recent report for the UK Government on the UK nuclear supply chain:

> 'The naval and civil reactor industries are often viewed as separate and to some extent unrelated from a government policy perspective. However, the timeline of the UK nuclear industry has clear interactions between the two, particularly from a supply chain development point of view.'[55]

So, important as it is, the debate over Trident may not be all that it seems. If this analysis is even partly correct, the stakes are even more extensive than the momentous issues that at first meet the eye. Bound up with the grave ethical, strategic, economic and political concerns that bear directly on the renewal of nuclear weapons capabilities, are a series of further evident questions around deeper forms of lock-in to nuclear technologies more generally. That these questions remain largely undiscussed in UK policy debates over either Trident or nuclear power, arguably constitutes one of the gravest implications of all – one that threatens not just the outcomes of policy making in either of these particular areas, but the very processes of democracy itself.

Notes

1 HM Government. National Security Strategy and Strategic Defence and Security Review 2015. London: The Stationary Office, Crown Copyright; 2015.
2 CND. People not Trident: the economic case against Trident. London: Campaign for Nuclear Disarmament; 2014.
3 Edwards R. Revealed: MoD's new multi-million pound Trident deal with America. The Herald Scotland online [Internet]. Edinburgh; 2014 Nov 23; Available from: http://www.heraldscotland.com/news/home-news/revealed-mods-new-multi-million-pound-trident-deal-with-america.25941179
4 Ritchie N. A nuclear weapons free world? Britain, Trident, and the challenges ahead. London: Palgrave Macmillan; 2012.
5 BASIC. Trident Replacement : The Facts. :1–6. 2013.
6 BASIC. US-UK Mutual Defence Agreement Renewal 2014: a foregone conclusion? [Internet]. British American Security Information Council Webpages. 2014 [cited 2016 Apr 24]. Available from: http://www.basicint.org/blogs/2014/02/us-uk-mutual-defence-agreement-renewal-2014-foregone-conclusion
7 Johnstone P, Stirling A. Why Germany is dumping nuclear power – and Britain isn't [Internet]. The Conversation. 2015 [cited 2016 Apr 23]. Available from: https://theconversation.com/why-germany-is-dumping-nuclear-power-and-britain-isnt-46359
8 Johnstone P, Stirling A. Comparing Nuclear Power Trajectories in Germany And the UK : From 'Regimes' to 'Democracies' in Sociotechnical Transitions and Discontinuities. SPRU Work Pap Ser [Internet]. 2015;18:1–86. Available from: www.sussex.ac.uk
9 SPRU. Governance of Discontinuity in Technological Systems (DiscGo) [Internet]. Science Policy Research Unit (SPRU), University of Sussex webpages. 2016 [cited 2016 Apr 27]. Available from: http://www.sussex.ac.uk/sussexenergygroup/research/current/discgo
10 House of Commons Defence Committee. The Defence Industrial Strategy: update. London: The Stationary Office; 2007.
11 The Herald. UK needs Trident to maintain 'outsized role on global stage', says

US. The Scottish Herald [Internet]. Edinburgh; 2016 Feb; Available from: http://www.heraldscotland.com/news/14275047.UK_needs_Trident_to_maint ain__outsized_role_on_global_stage___says_US/?ref=rss

12 Ireland G. The Challenges Facing the UK Submarine Base. Vol. 3. London; 2012.

13 Edwards R. MoD struggling with shortage of nuclear engineers [Internet]. The Herald Scotland online. 2015 [cited 2015 Mar 8]. Available from: http://www.heraldscotland.com/news/home-news/mod-struggling-with-shortage-of-nuclear-engineers.120130614

14 IAEA. IAEA Safeguards in Practice [Internet]. International Atomic Energy Agency Webpages. 2016 [cited 2016 Apr 27]. Available from: https://www.iaea.org/safeguards/safeguards-in-practice

15 IPFM. 2015 Fissile Materials Report: Nuclear Weapon and Fissile Material Stockpiles and Production. INternational Panel on Fissile Materials; 2015.

16 Bergeron K. Tritium on ice: The dangerous new alliance of nuclear weapons and nuclear power. Boston: MIT Press; 2002.

17 Arnold L. Britain and the H-Bomb. Basingstoke: Palgrave; 2001.

18 Sovacool BK, Valentine S V. The International Politics of Nuclear Power: Economics, Security, and Governance. Oxon: Routledge; 2012.

19 BBC Radio 4. The Today programme interview with Amber Rudd. UK: BBC; 2016.

20 Schneider M, Froggatt A. World nuclear industry status report 2015. Paris: A Mycle Schneider Consulting Project; 2015.

21 Randall T. Fossil fuels just lost the race against renewables [Internet]. Bloomberg. 2015 [cited 2016 Apr 21]. Available from: http://www.bloomberg.com/news/articles/2015-04-14/fossil-fuels-just-lost-the-race-against-renewables

22 Cameron A. Government U-turn on renewables shows gas, oil and nuclear are still favourites. The Guardian [Internet]. London and Manchester; 2015 Dec; Available from: http://www.theguardian.com/sustainable-business/2015/dec/20/government-u-turn-renewables-gas-oil-nuclear-favourites

23 DECC. Assessment SMRT. Early Market Engagement Announcement. 2015;(March).

24 Milmo D, Harvey F. Nuclear giants RWE and E.ON drop plans to build new UK reactors. The Guardian Online [Internet]. London and Manchester; 2012; Available from: http://www.theguardian.com/environment/2012/mar/29/nuclear-reactors-rwe-eon-energy

25 Wynn Kirby P. Europe's new nuclear experience casts a shadow over Hinkley. The Guardian Online [Internet]. London and Manchester; 2014 Mar; Available from: http://www.theguardian.com/environment/2014/mar/25/europes-new-nuclear-experience-casts-a-shadow-over-hinkley

26 Gosden E. Hinkley Point C: the story so far. The Daily Telegraph [Internet]. London; 2015 Mar; Available from: http://www.telegraph.co.uk/

news/earth/energy/nuclearpower/11404344/Hinkley-Point-new-nuclear-power-plant-the-story-so-far.html
27 Broomby R. Questions about UK scrutiny of Chinese nuclear tie-up [Internet]. BBC News Online. 2015 [cited 2015 Jan 15]. Available from: http://www.bbc.co.uk/news/uk-politics-30778427
28 Samuel H. New UK nuclear plants under threat as 'serious anomaly' with model found in France. The Daily Telegraph [Internet]. London; 2015 Apr; Available from: http://www.telegraph.co.uk/news/worldnews/europe/france/11546271/New-UK-nuclear-plants-under-threat-as-serious-anomaly-with-model-found-in-France.html
29 Macalister T. EDF Finance Minister resigns. The Guardian Online [Internet]. London and Manchester; 2016 Mar; Available from: http://www.theguardian.com/environment/2016/mar/07/hinkley-point-c-nuclear-project-in-crisis-as-edf-finance-director-resigns
30 The Week. Hinkley Point nuclear project could still be postponed. The Week Magazine Online [Internet]. 2016; Available from: http://www.theweek.co.uk/60778/hinkley-point-nuclear-project-could-still-be-postponed
31 DECC. DECC Contracts for Difference auction [Internet]. London; 2016. Available from: https://www.gov.uk/government/uploads/system/uploads/attachment_data/file/407059/Contracts_for_Difference_-_Auction_Results_-_Official_ Statistics.pdf
32 The UK Govenment's Taskforce on Sustainable Consumption. Decentralised Energy: Business Opportunity in Resource Efficiency and Carbon Management [Internet]. Cambridge: University of Cambridge; 2008. Available from: http://www.cisl.cam.ac.uk/publications/publication-pdfs/decentralised-energy.pdf
33 Brown P. Voodoo Economics and the doomed nuclear renaissance. London; 2008.
34 Birmingham Policy Comission. The Future of Nuclear Technology in the UK. Birmingham; 2012.
35 Environmental Audit Committee. Keeping the lights on: Nuclear, Renewables and Climate Change. Vol. I. London; 2006.
36 Political Resources. UK political party manifestos [Internet]. Political Resources webpages. 1997. Available from: http://www.politicsresources.net/area/uk/man/man97.htm
37 PIU. The Energy Review [Internet]. London; 2002. Available from: http://www.gci.org.uk/Documents/TheEnergyReview.pdf
38 DTI. Our Energy Future: Creating a Low Carbon Economy. London; 2003.
39 Taylor S. The fall and rise of nuclear power in Britain. Cambridge: UIT Cambridge; 2016.
40 Royal Courts of Justice. The Queen on the applications of Greenpeace limited Vs Secretary of State for Trade and Industry' [Internet]. London; 2007. Available from: http://www.greenpeace.org.uk/MultimediaFiles/
41 House of Commons Trade and Industry Select Committee. New nuclear?

Examining the issues. London: The Stationary Office; 2006.
42 BBC News. Blair defiant over nuclear plans [Internet]. BBC News Online. 2007 [cited 2007 Jan 12]. Available from: http://news.bbc.co.uk/1/hi/uk_politics/6366725.stm
43 Vidal J. New nuclear row as green groups pull out. The Guardian Online [Internet]. London and Manchester; 2007 Sep; Available from: http://www.theguardian.com/environment/2007/sep/07/nuclearindustry.nuclear power
44 BERR. A White Paper on Nuclear Power, Meeting the energy challenge. London; 2008.
45 KOFAC. About Us [Internet]. Naval Ship Building North West Webpages. 2016 [cited 2016 Apr 15]. Available from: http://www.navalshipbuilding.co.uk/navalship_home.asp?ID=HOMA
46 Schank JF, Riposo J, Birkler J, Chiesa J. THe United Kingdom's Submarine Industrial Base Volume 1: Sustaining Design and Production Resources. Pittsburgh: RAND Corporation; 2005.
47 Schank JF, Cook CR, Murphy R, Chiesa J, Pung H, Birkler J. The United Kingdom's Nuclear Submarine Industrial Base Volume 2: Ministry of Defence Roles and Required Technical Resources. Arlington: RAND Corporation; 2005.
48 Raman R, Murphy R, Smallman L, Schank JF, Birkler J, Chiesa J. The United Kingdom's Nuclear Submarine Industrial Base Volume 3: Options for Initial Fuelling. Arlington: RAND Corporation; 2005.
49 Ministry Of Defence. Defence Industrial Strategy: Defence White Paper [Internet]. London: The Stationary Office; 2005. 286-310 p. Available from: http://www.informaworld.com/openurl?genre=article&doi=10.1080/14702430802252545&magic=crossref
50 Stocker J. The United Kingdom and Nuclear Deterrence. Oxon: Routledge; 2007.
51 Ireland G. Beyond Artful : Government and Industry Roles in Britain's Future Submarine Design , Build and Support. The Royal United Services Institute Whitehall Report 3-07. London: Royal United Services Institute; 2007.
52 Cogent. Power People: The Civil Nuclear Workforce 2009-2025. Warrington; 2009.
53 Ministry Of Defence. The United Kingdom's Future Detterent Capability. London: The Stationary Office; 2008.
54 The Nuclear Institute. Nuclear Institute [Internet]. Nuclear Institute Web pages. 2016 [cited 2016 Apr 12]. Available from: http://www.nuclearinst.com/Homepage
55 Oxford Economics. The economic benefit of improving the UK's nuclear supply chain capabilities. Oxford: Oxford Economics Institute; 2013. 1-122

Nuclear Reactors and Climate Change

Pete Roche and Dr Ian Fairlie

This article first appeared at ianfairlie.org in September 2021 and is reproduced here with permission.

Can Small Modular Reactors and/or Advanced Nuclear Reactors Help Tackle Climate Change?

Summary

It is clear we need to tackle climate change quickly and effectively. With energy policies, several ways exist to mitigate carbon emissions and we need to compare them as to how rapid, how realistic and how cost-effective they are. In a few countries, nuclear proponents are lobbying, increasingly frantically, for new types of nuclear reactors to be constructed. But these are not even at their design stages, and many scientific analyses reveal that they are slow to implement and are hopelessly costly both in terms of their construction costs and eventual would-be electricity prices. On the other hand, renewable energy and energy efficiency programmes are here and now, are inexpensive, can be implemented quickly and do not have the myriad of problems associated with nuclear projects. Investment in costly nuclear power programmes, which would take decades to implement, would effectively worsen climate change because each pound spent on nuclear would be buying less solution which won't save carbon until it's much too late.[1]

Introduction

The world's leading climate scientists on the UN's Intergovernmental Panel on Climate Change (IPCC) have warned that we have fewer than 10 years to make massive and unprecedented changes to global energy infrastructure in order to limit global warming to moderate levels.[2] Even the most optimistic projections don't foresee any new reactor designs coming on stream until the 2030s and 2040s, and it would be even later before significant amounts of electricity were produced.

For example, Dr Gregory Jaczko, former chairman of the U.S. Nuclear Regulatory Commission (2009–2012) says we should only support nuclear projects

> "if they can compete with renewables and storage on deployment cost and speed, public safety, waste disposal, operational flexibility and global security. There are none [that can do that] today".[3]

What are SMRs and Advanced Nuclear Technologies?

Over the past few years, in a few countries with nuclear power programmes, (ie UK, Canada and the US), nuclear lobbyists have pressed for 'small' modular reactors' (SMRs), along with so-called "advanced" nuclear technologies (ANTs). These advocates allege that such nuclear projects could provide 'low carbon' energy solutions, although the large carbon arisings from uranium mining and milling and nuclear wastes are usually ignored in such claims.

It should be noted that many other countries (ie Germany, Austria and most EU countries including Ireland) have failed to support such claims. Indeed, the EU's taxonomy process, which is setting guidelines to apply to future EU support for energy projects, has pointedly refused to include nuclear projects as they are unsustainable both in environmental and economic terms.

Instead, independent commentators suggest that the nuclear industry and its protagonists are making these unsupported claims in order to stop the nuclear industry's actual (and apparently terminal) decline throughout the world, as many nuclear reactors are closing down at the ends of their lives.

SMRs and ANTs

The terminology used by nuclear supporters is unfortunately confusing. The UK government uses the term 'Advanced Nuclear Technologies' (ANTs) to cover two broad categories based on their technologies. First are reactors based on the same technology as existing reactors – ie Small Modular Reactors (SMRs).

Second are proposed Advanced Modular Reactors (AMRs) which have never operated successfully anywhere in the world.

- helium-cooled graphite-moderated high-temperature reactors (HTGR);

- sodium-cooled fast reactors (FBR);
- molten salt reactors;
- lead-cooled fast reactors.[4]

However other supporters lump the two categories under one heading of SMRs[5] or small, modular light-water reactors, and non-light-water "advanced" reactors.

The usually unstated reason for the nuclear industry's promotion of SMRs and AMRs is that existing large nuclear reactors are now uneconomic: some are even being shut because they are unprofitable to operate even after their capital costs have long been paid off. More important, new large reactors are exceedingly expensive to construct.

Therefore SMRs are being promoted as a solution to the high operating costs and to the difficulties of financing larger reactors. But the reason why existing reactors are large was precisely to derive economies of scale: why smaller reactors should be more economic is problematic. Nuclear proponents allege that assembly-line technology will be used in reactor construction but this has yet to be shown in practice anywhere in the world.

In addition, for a company to be confident enough to invest in a factory to manufacture reactors, it would need to ensure a market exists for them, and it would need to build a massive supply chain since none of it currently exists. Funding for that would presumably come from customer orders. But those customers are unlikely to appear until the designs and costs have been proven.

Other major obstacles remain. Some are technical, some are regulatory, and some are due to the resistance by local groups to having nuclear reactors in the midst of their communities. And the financing of such schemes would only be possible with significant subsidy from taxpayers. In view of these manifest problems, some say that SMRs are little more than wishful thinking. For example, Professor MV Ramana – Simons Chair in Disarmament, Global and Human Security at the School of Public Policy and Global Affairs at the University of British Columbia – states:

> "SMR proponents argue that they can make up for the lost economies of scale by savings through mass manufacture in factories and resultant learning. But, to achieve such savings, these reactors have to be manufactured by the thousands, even under very optimistic assumptions about rates of learning."[6]

And Dr Gregory Jaczko agrees. "Only wide-scale adoption of the technology would deliver those benefits and there is no obvious market to support that today."

UK Rolls Royce

In the UK, Rolls Royce is promoting a 450 megawatts (MW) reactor. But there is confusion about whether this is "small" or large and what is meant by this adjective in the name "small modular reactor". SMRs are generally expected to have a capacity of less than 300 MW compared with the 800 MW capacity of the reactors being built at Hinkley Point C in Somerset, England. Strangely, the reactor currently being promoted by Rolls Royce in the UK is larger than most of the UK's now closed Magnox reactors, and very similar in size to the UK's existing AGR reactors.

Rolls Royce claims its reactors could cost as little as £2bn each, and says its first SMR could be operating in the 2030s.[7] The company says it plans to build 16 SMRs in the UK by 2050. A consortium led by Rolls Royce says it has secured at least £210m needed to unlock a matching amount of taxpayer funding, so that it can submit its SMR design to the nuclear regulators for approval.[8] Rolls-Royce claims that its SMRs could generate power at a cost of £60/MWh[9]. But several commentators say these estimates are implausible and far too small. In addition, Rolls Royce is demanding significant UK government funding to pursue its project and is threatening to abandon it if government largesse is not forthcoming.[10]

Dr. Gregory Jaczko says; "...the nuclear industry always promises better, faster and cheaper yet it fails to deliver ... Small modular designs are only promising to be cheaper than traditional reactors. Current estimates show they are more expensive than renewables, like wind and solar, even with storage and without subsidies. Small reactors have a long way to go to be competitive. Dramatic cost decreases for high-volume energy storage, which address the intermittency of some renewables, make the competitive case for any form of nuclear even tougher."

Advanced Modular Reactors

The second category of Advanced Modular Reactors, ie. non-light-water "advanced" reactors are even more pie-in-the-sky than SMRs. AMRs are largely based on notoriously unsuccessful concepts from more than 50 years ago. They remain unproven today.

Unlike light-water reactors, these designs rely on materials other than water for cooling. Some developers contend that these reactors, still in the concept stage, will solve the problems that have plagued light-water reactors and be construction-ready by the end of this decade.[11] However, a

Union of Concerned Scientists (UCS) analysis in the US suggests that this outcome may be just as likely as electricity being "too cheap to meter." Written by UCS physicist Dr Edwin Lyman, the 140-page report found that these designs are no better — and in some respects significantly worse — than the light-water reactors in operation today.[12]

Lyman took a close look at the three main designs here: sodium-cooled fast reactors, high-temperature gas-cooled reactors and molten salt–fuelled reactors. Many developers maintain, with little or no hard evidence, they will be cheaper, safer and more secure than currently operating reactors; will burn uranium fuel more efficiently, produce less radioactive waste, and reduce the risk of nuclear proliferation; and could be commercialized relatively soon.

Those claims do not hold up to even elementary levels of scrutiny. One of the sodium-cooled fast reactors, TerraPower's 345-megawatt Natrium, has received considerable media attention because it is supported by billionaire Bill Gates. But a massive problem associated with sodium-cooled reactors is the use of molten sodium itself. This burns fiercely when exposed to air and explodes when exposed to water. The disastrous experiences of the UK's Dounreay Fast Reactor and Japan's Monju reactor attest to the severe problems with liquid sodium. Lyman at UCS also believes the Natrium's design could experience uncontrollable power increases that would result in rapid core melting.

In an open letter to Bill Gates, Arnie Gundersen, former nuclear operator and now Chief Engineer of Fairewinds Energy Education says he fears:

> "you have made an enormous mistake by proposing to build a sodium-cooled Small Modular Reactor (SMR) in Wyoming … your atomic power company Natrium (the Latin word for sodium), is following in the footsteps of a seventy-year-long record of sodium-cooled nuclear technological failures. Your plan to recycle those failures and resurrect liquid sodium again will siphon valuable public funds and research from inexpensive and proven renewable energy alternatives. Moreover, spending public funds on Natrium will make the global climate crisis worse, not better!"[13]

Dr Edwin Lyman concludes:

> "Unfortunately, proponents of these non-light-water reactor designs are hyping them as a climate solution and downplaying their safety risks. Given that it should take at least two decades to commercialize any new nuclear reactor

technology if done properly, the non-light-water concepts we reviewed do not offer a near-term solution and could only offer a long-term one if their safety and security risks are adequately addressed." Any federal appropriations for research, development and deployment of these reactor designs, he says, "should be guided by a realistic assessment of the likely societal benefits that would result from investing billions of taxpayer dollars, not based on wishful thinking."

Dr. Gregory Jaczko has added that even if these risks of electricity from small and advanced reactors were addressed, proliferation concerns and waste management problems would still be hurdles.

Thorium

Thorium has been mooted as a fuel in thorium reactors for many decades, but their past records (in the US and USSR) have been dismal. In addition, spent thorium fuel is a proliferation hazard. Strictly speaking, thorium fuel does not exist, since thorium-232 is not fissile, but it is fertile. When blended with fissile plutonium-239, both are used to fuel a nuclear reactor. Plutonium keeps the chain reaction going, and while that is happening, thorium-232 absorbs neutrons and is changed into uranium-233 which is fissile.[14] This is a severe proliferation hazard as isotopically pure uranium 233 is suitable for making nuclear weapons. Therefore spent thorium fuel would be a tempting target for theft by terrorists.[15]

Robert Alvarez, former senior policy adviser to the secretary and deputy assistant secretary for national security and the environment of the US Department of Energy, says the United States tried to develop thorium as an energy source for some 50 years with no success. Sadly it is still struggling to deal with the legacy of those attempts. In addition to the $ billions it spent fruitlessly to develop thorium fuels, the US government will have to spend billions more, at numerous federal nuclear sites, to deal with the wastes produced by those efforts.[16]

Conclusions

Even in the extremely unlikely event that some of the claims of Advanced Nuclear Reactor proponents proved to be correct, building a sufficient number of these reactors to make any impact on carbon emissions would take far too long. We simply do not have the time to do this.

In the meantime, expending time, money and efforts on these unproven reactor dreams is a dangerous distraction from implementing more effective climate mitigation programmes. Renewable energy exists and is cheap and becoming cheaper, and needs little or no public subsidy - a big contrast with new nuclear. Many energy efficiency schemes can actually be implemented at negative net cost.

Many studies now show that it is perfectly feasible to run energy systems using 100% renewable energy in many countries and regions. See the abstracts of 56 peer-reviewed published articles from 18 independent research groups (with 109 authors) worldwide supporting the result that energy for electricity, transportation, building heating/cooling, and/or industry can be supplied reliably with 100% or near-100% renewable energy at difference locations worldwide.[17]

Many nuclear advocates call for a 'balanced energy policy' and promote the idea that 'we need every energy technology' in order to successfully tackle climate change. Of course, implicit here is the need for some nuclear capacity.

But these calls suggest we have infinite amounts of money to spend on energy projects. We do not: resources are scarce and we need to make choices.

Because climate change is a serious and urgent problem then we must spend our limited resources as effectively as possible on projects which can deliver carbon reductions as quickly as possible. Investment in untried, untimely and expensive nuclear power would, in effect, worsen climate change because each pound spent on nuclear is buying less solution than it would do if we were to spend it on energy efficiency and renewables.[18]

References

1. *Forbes* 18th November 2019
https://www.forbes.com/sites/amorylovins/2019/11/18/does-nuclearpower-slow-or-speed-climate-change/?sh=2ed39ab7506b

2. *Guardian* 8th October 2018
https://www.theguardian.com/environment/2018/oct/08/global-warmingmust-not-exceed-15c-warns-landmark-un-repor

3. *The Hill* 23rd Feb 2021 https://thehill.com/opinion/energy-environment/539991-climate-change-and-advanced-nuclear-solutions

4. Advanced Nuclear Technologies 10th July 2020
https://www.gov.uk/government/publications/advancednuclear-technologies/advanced-nuclear-technologies

5. New Nuclear Monitor 7th March 2019
https://wiseinternational.org/sites/default/files/NM872-873-final.pdf

6. ibid see page 12

7. *Telegraph* 5th August 2021 https://www.telegraph.co.uk/business/2021/08/05/rolls-royce-edges-back-profit-pushes-recovery-forecasts/

8. *Telegraph* 3rd August 2021 https://www.telegraph.co.uk/business/2021/08/03/rolls-royce-lines-funding-mini-nuclear-reactor-revolution/

9. Reuters 7th May 2018 https://www.reutersevents.com/nuclear/rolls-royce-smr-use-site-factories-hit-60-poundsmwh

10. New Nuclear Monitor 7th March 2019 https://wiseinternational.org/sites/default/files/NM872-873-final.pdf

And Prospects for Small Modular Reactors by Steve Thomas, Paul Dorfman, Sean Morris and MV Ramana. July 2019 https://www.nuclearpolicy.info/wp-content/uploads/2019/07/Prospects-for-SMRs-report-2.pdf

11. *Scientific American* 23rd July 2021 https://www.scientificamerican.com/article/lsquo-advanced-rsquo-nuclear-reactors-don-rsquo-t-hold-your-breath/

12. Advanced Isn't Always Best by Ed Lyman, Union of Concerned Scientists, 18th March 2021 https://www.ucsusa.org/resources/advanced-isnt-always-better

13. *Counterpunch* 20th August 2021 https://www.counterpunch.org/2021/08/20/an-open-letter-to-bill-gates-about-his-wyoming-atomic-reactor/

14. *Times* 16th Sept 2021 https://www.thetimes.co.uk/article/experimental-reactor-could-hand-china-the-holy-grail-of-nuclear-energy-tcsqxwp3m

15. *Bulletin of Atomic Scientists* 6th August 2018 https://thebulletin.org/2018/08/thorium-power-has-a-protactinium-problem/

16. *Bulletin of Atomic Scientists* 11th May 2014 https://thebulletin.org/2014/05/thorium-the-wonder-fuel-that-wasnt/

17. Abstracts of 56 Peer-Reviewed Published Journal Articles From 18 Independent Research Groups With 109 Different Authors Supporting the Result That Energy for Electricity, Transportation, Building Heating/Cooling, and/or Industry can be Supplied Reliably with 100% or Near-100% Renewable Energy at Difference Locations Worldwide April 7, 2021 https://www.no2nuclearpower.org.uk/wp-content/uploads/2021/04/100PercentPaperAbstracts-2021.pdf

18. *Forbes* 18th November 2019 https://www.forbes.com/sites/amorylovins/2019/11/18/does-nuclear-power-slow-or-speed-climate-change/?sh=2ed39ab7506b

* * *

Nuclear Fusion – Not The Answer To Our Energy Needs
Ian Fairlie (ianfairlie.org)
September 12, 2021

The government is proposing a fusion reactor – the Spherical Tokamak for Energy Production (STEP) – for our energy needs. This would be a smaller version of the unsuccessful Tokamak prototype (JET) at Culham in Oxfordshire. Quite why the STEP project would be expected to work when its prototype has failed is unexplained in official documents. But the government is looking to develop an operational site for a STEP reactor. The plan is for the Business Secretary to choose a site for a prototype, following recommendations of UKAEA, by 2024.

However the government's new panacea has almost nothing to do with our energy needs and everything to do with Boris Johnson's ill-considered techno-dreams. It will most likely join the long line of Boris' flops after the Thames Gateway airport, the Emirates airline cable car, the bendy bus, the Thames Garden bridge etc, etc. But this time the taxpayer will have to pay £billions rather than £millions.

What is nuclear fusion?

It's a dangerous process whereby radioactive hydrogen (tritium) is smashed into another form of hydrogen (deuterium) at massive temperatures and pressures inside a plasma to release much radiation and some heat. The same process occurs in our Sun ….. but the Sun is safely located 93 million miles away.

Formidable technical problems exist with fusion. First they have to get the deuterium-tritium reaction to work continuously: they've done this at the JET experimental facility at Culham... for a few seconds. Then they have to get it to release more energy than used in producing the reaction. This has never successfully happened to date. Then they would have to capture the energy released. JET has never been close.

The plan is to surround the plasma chamber with molten lithium. But the engineering is really invidious: a high vacuum on one side, molten lithium on the other, and billions of high-energy neutrons bombarding the wall each second. They then have to run hot molten lithium through heat exchangers to raise steam for a turbine. Experience with such heat exchangers – molten metal on one side, water on the other – has been disastrous all over the world. The problem is that lithium is extremely flammable, indeed explosive in contact with water or air. And should it ever operate, vast amounts of radioactive gases and radioactive water vapour would be released to the local environment.

The government's mooted fusion reactor comes with promises of cheap and clean energy to move to a zero-carbon economy, with little radioactive waste and no plutonium by-products for nuclear weapons. But this government has a bad track record with its promises ... how valid are these claims?

The reality is that a fusion reactor, if ever operated, would produce many radioactive by-products that are far from harmless. In addition, most (around 80%) of the output energy would be in the form of high-energy neutrons which would lead to structural damage, large amounts of radioactive waste and the need for much biological shielding to protect operators and the public nearby.

Fusion plants can also be viewed as gigantic exercises in tritium recycling. If the plant were ever constructed, large amounts of radioactive tritium (~1018 becquerels per year) would be routinely released into the atmosphere and via the cooling water. That's 1,000,000,000,000,000,000 Bq each year – a great deal of radioactivity. It would contaminate all areas downwind and downstream. Some nuclear scientists think that tritium is a "weak" nuclide but the reality is the opposite: see *The Hazards of Tritium* – Dr Ian Fairlie. If an explosion and/or fire occurred (tritium and deuterium are both flammable), the amounts of radioactivity released would be even greater and would constitute a nuclear disaster.

Fusion reactors would also be subject to most of the major problems associated with fission reactors, including large-scale cooling demands, high construction and operational costs and lengthy construction times – stretching to decades. The structure, damaged by neutron bombardment, would need to be replaced regularly, resulting in large amounts of radioactive wastes for which there is no current solution in the UK.

What do experts say?

In the past, skeptical scientists opposed nuclear fusion as unworkable, including many US scientists. More recently, Dr Daniel Jassby who worked for 25 years on plasma physics and neutron production related to fusion energy at the Princeton Plasma Physics Laboratory, at Princeton university in the US has written two informative articles (see below) on the myriad problems with nuclear fusion for the *Bulletin of Atomic Scientists* (the US journal which gave us the Doomsday Clock). He concluded "When you consider we get solar and wind energy for free, to rely on fusion reaction would be foolish".

"Fusion reactors: Not what they're cracked up to be" – by Daniel Jassby, April 19, 2017: https://thebulletin.org/2017/04/fusion-reactors-not-what-theyre-cracked-up-to-be/

"ITER is a showcase ... for the drawbacks of fusion energy" – by Daniel Jassby, February 14, 2018: https://thebulletin.org/2018/02/iter-is-a-showcase-for-the-drawbacks-of-fusion-energy/

In short, nuclear fusion would not provide cheap, clean, safe or healthy energy and would reduce the funding available for safer and cheaper renewable energies.

Stop Trying to Make Nuclear Power Happen

Dave Cullen

Dave Cullen researches nuclear weapons for a living. This article was first published by New Socialist *(newsocialist.org.uk) in October 2021.*

Nuclear power used to be a big thing: a huge touchstone issue on which people across the political spectrum based part of their identity. If you knew where you stood on the left-right spectrum in the '80s, your position on nuclear power followed naturally. That is no longer the case.

Some remember where they stood, and are in no mood to forgive and forget. Others are a bit embarrassed about their former zealotry. A handful of people with a surface understanding of the issue have decided to define themselves against their strident forebears and paint themselves as reasonable pragmatists. Unfortunately, a handful of prominent public environmentalists are members of this last group.

Their position is particularly unfortunate because of the poverty of public debate on this issue. It would not be so harmful if these people, who mostly reached adulthood after the battlelines on this issue had already been drawn, were airing what should be understood as their contrarian views within a healthy fact-based discourse. But instead they are parroting a PR line from the nuclear power industry in a context where almost everyone is operating from a position of little to no understanding.

In that context, it's completely understandable that some comrades have also accepted the industry line that nuclear power is necessary to tackle climate change. I think this is a huge mistake, for the reasons I will set out in this piece, but I have no interest in picking a fight with anyone who formed that belief in good faith. Although I think nuclear power is a fundamentally bad idea that nobody on the left should have any truck with, supporting nuclear power is not like supporting migrant detention or abstaining on the spy cops bill. However, I do think a lot of comrades need to think a lot more critically about whose interests are being served, and what answers are predetermined by framing the issue in terms of choosing between nuclear power and the greater evil of climate change.

Let's be clear: nuclear power is not worse than climate change. But that doesn't matter. Comparing the two is a pointless false dichotomy, and the focus of all the public debate in the UK on this framing is the main reason nuclear power is being built here at all. In fact, nuclear power is an expensive dead end, and pursuing it will make climate change worse. Furthermore, nuclear power has several unalterable characteristics that mean it is fundamentally incompatible with the world we want, and need, to build.

Before we get onto why nuclear power should be anathema to anyone on the left, we first need to spend a bit of time looking at the current situation and the history of nuclear power to understand how it has failed even on its own terms. In this piece, I'm not going to talk about the unique dangers of nuclear power or argue that the risks of radiation are often understated. That's not because there's no merit to those claims, it's because there's no way to tackle them properly in any piece shorter than book-length, and nuclear power's other flaws are sufficient without venturing into that territory.

The current state of nuclear new build in the UK

How then has nuclear power failed on its own terms? Let's start with the current situation. The UK is 15 years into a "nuclear renaissance". No reactors have been built during that time. No reactors have been built in the UK in over 25 years. One EPR reactor at Hinkley Point C has been under construction since around 2014, and will not be ready until at least mid-2026. Of the six new nuclear power stations that did recently stand a realistic chance of being constructed, three have been cancelled since late 2018. We're in the throes of a planetary emergency, and we need to decarbonise the electricity system over the next decade.[1] If nuclear power is going to play any serious role in that process the next few years need to be radically different for the industry. They won't be.

Hinkley C is now predicted to start producing power at least three years later than its original 2023 completion date, and a year beyond the date given when the Cameron government saved the project by guaranteeing EDF would be paid for the power the plant produces at double the market rate. The estimated cost of the project has risen from £16bn to £23bn.

Hinkley C is actually going quite well compared to earlier EPR projects. The first two EPR reactors to be started at Olkiluoto in Finland and Flamanville in France have both been under construction for more than 14 years and are expected to cost three and five times their original budgets respectively.

There are a couple of EPR reactors built in China that are actually producing power[2], so EDF will probably eventually complete the two planned Hinkley reactors – if only because the company will probably go under if it doesn't, and the French government won't allow that. EDF has public plans to build more EPR reactors at Sizewell and Moorside, though these plans primarily serve to make the company look like a healthy proposition. In reality it's in no position to deliver on them.

EDF is substantially in debt, and its total liabilities for decommissioning its French nuclear fleet are almost twice the size of the company's market capitalisation.[3] It is solvent on paper because of the effect of discounting.[4] One of the reasons for its €50bn project to extend the life of its French reactor fleet from 40 years to 50 is to delay those decommissioning costs for another decade. The French state auditor has said EDF can't afford to build any more EPRs in France, and shouldn't unless someone else puts up the money. Construction at Hinkley is being financed through EDF's balance sheet rather than the low-interest loans offered by the UK Treasury, because these loans were contingent on the reactor in Falmanville being ready by 2020. This means Hinkley won't actually be that profitable for EDF because the cost of borrowing to finance construction will be much higher than originally planned.

Despite its resemblance to a doomed pyramid scheme, EDF is still the most promising company in the UK's supposed nuclear revival. NuGeneration, which was going to build Moorside just across the road from Sellafield, died when its owner Westinghouse went bankrupt. Westinghouse's fate gives an indication of where EDF might have ended up without the backing of the French state. Losses from a failed attempt to build two AP1000 reactors in South Carolina sunk the company so comprehensively that it nearly bankrupted its parent company Toshiba – forced to sell its profitable microchip subsidiary in order to offset Westinghouse's catastrophic losses. The EPR and AP1000 were the big hope for the European and North American nuclear industries, the cream of the so-called advanced 3rd generation reactors.[5] They were supposed to address some of the shortcomings of earlier generations of reactors by, for example, using modular designs to reduce upfront costs and construction time. Instead they stand a good chance of being the industry's epitaph.

Horizon, who were hoping to build new reactors at Wylfa and Oldbury, scrapped their projects after the government failed to buckle and underwrite the projects using a funding mechanism called "Regulated Asset Base" (RAB). RAB means that current users of electricity would fund the project through a surcharge on their bills, even though the power station wouldn't be producing any electricity for a decade or so. The government had offered to take a $1/3$ stake in Horizon, consider financing debt for building the reactors and guaranteeing a fixed price for electricity similar to the deal for Hinkley, but Horizon were still unwilling to take the risk. They "suspended" the projects in early 2019, fired most of their staff, and then finally withdrew in September 2020.

The only other serious player in the UK is the Chinese state-owned CGN, who are hoping to build a HPR1000 reactor at Bradwell. The opportunity to build this reactor was offered as an incentive to get CGN to invest in Hinkley C. Before they can apply for a site licence to build Bradwell, the HPR1000 reactor design needs to complete a generic assessment, a process that will not be completed before 2022.[6] I don't know whether the regulator will approve the HPR1000, but I think there's a pretty decent chance that CGN's Bradwell plans will be blocked. CGN is subject to US export controls and, after the controversy over Chinese firms and the 5G network, it's difficult to imagine a state firm from a country the government has just officially deemed a 'systemic challenge [...] to our security, prosperity and values' being allowed to run a nuclear power station in Essex.

These are the entities who are supposedly poised to roll out a massive expansion of nuclear power in the UK. There are other companies whose PR agents would like you to believe that they are poised to start building new reactors. It's important to understand that these companies are not able to build nuclear power stations. They either do not have the capital or they do not want to take the risk. What they are trying to achieve with the regular fluff pieces they place in the press is a sense of momentum. The end goal is either for someone with more money than sense to buy their patents or the company as a whole, or, more likely, for credulous MPs to lobby the government to subsidise their schemes. As Mycle Schneider, editor of the annual *World Nuclear Industry Status Report*, has said, "If the industry doesn't launch phantom projects, then it will die even faster."

"Phantom projects" is the best description of SMRs, subject of the industry's current PR blitz and therefore favoured talking point of a particular kind of nuclear bore. Ignore the funding the government has announced for them. There is no operator in the UK market that is currently in a position to deploy them, and the funding is best understood as a hidden subsidy for Rolls-Royce, who build reactors for UK nuclear submarines and want SMRs to support this aspect of their business.

There are no SMR designs that have been submitted for generic approval from the regulator. SMRs are never going to be deployed at scale in the time we have available to tackle climate change. There are really basic engineering reasons to believe that they will actually be more expensive than larger nuclear power stations, rather than cheaper, and nobody has come up with a convincing argument to explain how having lots of smaller reactors and the subsequent waste close to cities is going to be acceptable in terms of safety and security.

SMRs are just the latest spin that the industry is only one small step away from breaking free from its fundamental flaws and finally ushering in the long-promised era of electricity that is too cheap to meter. Before SMRs, it was "recycling" nuclear waste to make mixed oxide fuel. That gave us the disastrous THORP reprocessing plant that will probably end up costing more money to build and dismantle than it ever made in contracts, and has left the UK sitting on nearly 140 tons of civil plutonium, the world's largest stockpile.

The Beginnings of the 'Renaissance'

As with so many rotten and depressing aspects of British life in 2021, a direct causal line can be traced back from the current situation to decisions taken under the Blair government. In 2003, under Patricia Hewett, the Department of Trade and Industry published a White Paper. The White Paper was notable for being the first energy policy to try and tackle climate change, and because it said nuclear power was uneconomic. Following intense lobbying from the nuclear industry, which just happened to coincide with plans to replace the UK's nuclear armed submarines and build a new warhead, Downing Street stepped in and ordered a review that resurrected nuclear power in the UK. Before the review had even formally concluded, Blair was telling the CBI that nuclear power was "back on the agenda with a vengeance". A High Court judge later ruled that review was unlawful and 'seriously flawed', but Blair got his way, and new nuclear power has been official government policy ever since.[7]

At the time of the 2003 White Paper, nuclear power was dead. There hadn't been a new reactor built since Sizewell C was completed in 1995. The supposedly most profitable part of the industry had been hived off and privatised in 1996. Within six years it was facing bankruptcy and the government had to step in to rescue it, picking up more of its decommissioning liabilities and pumping money into the company. The company was then re-privatised and sold on to EDF. The less profitable part of the industry (which includes several sites where the always nominal distinction between nuclear power and nuclear weapons production was completely nonexistent) stayed in government hands throughout and the cost of cleaning up the sites is expected to cost us somewhere between £99bn and £232bn. This is the industry that Blair decided to resurrect.

At the time, the official line was that there would be no government subsidy for nuclear, but everyone paying attention knew that was bullshit. No nuclear power station has ever been built without government support anywhere in the world, but the industry wasn't worried. The intention was for the government to commit so fully to nuclear that it would be left with no option other than to provide the necessary support. Even when the initial regulatory framework was being set out, there were clear government subsidies in the form of the government bearing most economic risks from an accident[8], and the costs of dealing with waste.[9] Despite this, and as many of us were predicting a decade ago, the government has since blinked and these subsidies have been followed by price guarantees and government-backed cheap finance, because nuclear power cannot be built without state support of this kind.

The reasons for this are straightforward. Nuclear requires a massive investment of upfront time and money. Constructing a reactor takes about a decade, and is an incredibly complex project requiring a highly specialist workforce. Running a reactor is comparatively easy, and barring any unforeseen problems you do get a supply of relatively cheap electricity for several decades. But after that decommissioning a reactor takes decades, and is again a highly complex project requiring substantial sums of money and human labour. The whole process generates wastes that have to be managed safely for millenia – opinions differ about what's involved in this last stage because nobody has ever opened a facility for high level nuclear waste, nor grappled with the millenia that follow.

The economics of nuclear power are therefore highly sensitive to fluctuations in the market price for electricity and the cost of finance, and only add up if the private operator can sell enough electricity for a high enough price during the years for operation to offset its construction, decommissioning and waste costs. In practice this is usually achieved by offloading as many of those costs as possible onto governments. However, the cost of renewables has fallen so quickly over the time it has taken the nuclear industry to half-build Hinkley C that even with much of the costs offloaded nuclear just isn't economic any more.

Nuclear power cannot tackle climate change

Why does any of this matter to us? So what if nuclear power has struggled under the conditions of a neoliberal energy market? Why should those of us who want to abolish all neoliberal market conditions care about that? I've introduced this potted history for several reasons. Firstly, to demonstrate that our public discourse on nuclear power is almost entirely bullshit. The industry and their numerous shills are peddling a fantasy and dictating terms of debate that bear almost no relation to reality. Secondly, because those struggles are not a feature of the market, they are a feature of nuclear power's intrinsic characteristics. But most importantly, because nuclear power's struggles in the market are fundamentally related to its unsuitability as a climate change mitigation technology.

Even with a hypothetical socialist government, under utopian conditions, prepared to properly resource this technology and find technical solutions to all the problems that crop up, nuclear power would still be extremely expensive, complex and slow to build. The planetary emergency that is climate change requires us to deploy technological solutions (alongside fundamental political, economic and social changes) that are relatively cheap, quick to put in place, scalable and available worldwide.

If we really did face a choice between nuclear power and climate change, we might need to accommodate ourselves to those shortcomings. But we don't, not because climate change isn't an issue, but because nuclear is inadequate to its threat. The lead-in time for nuclear power already rules it out if we are serious about adopting emissions scenarios that stand any chance of avoiding dangerous levels of warming. It doesn't make any sense at all to commit the vast sums nuclear requires when spending the same amount of money on renewables would provide almost five times as much generation capacity in a fraction of the time, particularly with nuclear's record of delays and failures.[10] Ploughing resources and political commitment into nuclear will only slow down the response to climate change, and trying to engineer around the myriad problems of nuclear power is a distraction we cannot afford from the engineering problems that we should be focussing on: using demand management, energy efficiency, storage, international interconnectors and non-polluting load-following power sources to make a grid powered by renewables a reality.

Nuclear isn't scalable to anything like the degree that would be needed for it to play a role in decarbonising electricity production worldwide. The state of the companies with a stake in the UK's nuclear newbuild scene says it all: a roll-call of hucksters and the technically insolvent. The institutional capacity to roll out nuclear power across the globe at scale is simply not there. These projects have massive and complicated supply chains with a handful of specialist companies in key positions. Recent scandals over faulty welding and fraudulent certification of steel in nuclear suppliers illustrate the degree of specialisation needed in the nuclear supply chain and the difficulties there would be in trying to bring in new suppliers in order to scale the industry up. The industry already has problems with quality control, and you cannot safely mass produce reactors on the cheap. Skilled staff is another huge bottleneck – nuclear engineers take at least four years to go through an entry-level qualification, and will need many years of experience before they are ready to be in a supervisory role. Even with the UK's current plans there is a crisis-level shortfall in skilled staff. There is no way that nuclear power can play a major role in the global energy transition that we need to see.

The complexity of nuclear power doesn't just create a prohibitive upfront cost – it also means that reactors are subject to unplanned outages that can last for years. When a turbine component in a wind farm breaks, the rest of the turbines continue to function as normal. When something goes wrong in a reactor, the only safe option is often to shut the whole reactor down, sometimes for months at a time.[11] Climate change also poses challenges for the siting of nuclear power plants. Reactors also need a constant supply of cold water for cooling, and during the heatwaves that are becoming more common under climate change, river water often gets too warm, an ongoing problem in France. The more common approach of siting reactors next to the sea is also problematic, however, as those sites will be threatened by sea-level rises and storm surges as climate change worsens. This is particularly an issue as some of those sites will probably need to store radioactive waste for decades after the plant shuts down.

Nuclear isn't even particularly suitable for filling the niche that its advocates have claimed for it: as a complement to renewables in a zero-carbon energy mix. A grid with lots of renewable power doesn't need a supply of always-on 'baseload' power, it needs flexible power sources to balance the variability of solar and wind. Nuclear could be used for this purpose, but the economics of nuclear power depend upon producing as much power as possible during the years of operation. Running a reactor below capacity in order to match variable power sources just makes it an even less financially viable option. Nuclear isn't even particularly suitable for filling the niche that its advocates have claimed for it: as a complement to renewables in a zero-carbon energy mix.

Nuclear Power and the Left

These are all reasons why any technocratic liberal with a grasp of the facts should shun nuclear power, but what about us on the left? I think there's a strand of thinking on the left that has been far too willing to accept the lies of the nuclear industry because it believes in technological progress as a general human good, and that we can solve a lot of social problems through technology. This is a good example of what John Bellamy Foster termed an "eco-modernist" position, where The Earth System crisis is said not to demand fundamental changes in social relations and in the human metabolism with nature. Rather it is to be approached in instrumentalist terms as a formidable barrier to be overcome by means of extreme technology.

In part, eco-modernism is an understandable reaction to the stereotypical Green belief that technological progress is actually bad. But we really need to move beyond approaching technology with a default attitude of either credulity or suspicion. Instead, let's consider individual technologies and the types of social relations and state forms that those technologies engender in order to decide whether they have a role to play in the future we need to build. A technology does not on its own cause social relations or forms of the state. Raymond Williams was right to insist, in their debate over the politics of nuclear disarmament, that E. P. Thompson's technologically determinist argument that nuclear weapons immediately give us a particular social and international order obscures decisive questions, including around the relationships that produced the technological form: "behind it, of course, is another question: who 'gave us' the hand-mill, the steam-mill, the missile factories?" However, Williams' refusal of the intellectual closure of Thompson's argument does not entail presenting technologies as politically indeterminate, as many eco-modernists would.

It is not possible to abstract nuclear power from its current context and purposes and simply transfer it to a socialist context and purposes. Nuclear power was, in Williams' terms, "consciously sought and developed" within particular social and international relationships, and features of the technology favour the maintenance of these relationships, and particular forms of the state. Considering nuclear power in this way, it's clear that its characteristics militate against the world we want to see.

Nuclear power requires large and secretive states and companies. The fundamental role of technologies and knowledge that could be used to create nuclear weapons, and the extensive upfront costs, makes state intervention in alliance with big capital, without any possibility of democratic planning, almost inevitable. The disparities of knowledge and of financial power that flow from these basic facts mean that nuclear power is inherently hierarchical and cannot be subject to meaningful democratic control. Unsurprisingly, given the overlap between the two technologies, many of the characteristics of Elaine Scarry's conception of a *Thermonuclear Monarchy*[12], which argues that nuclear weapons structurally require forms of secrecy and unaccountable powers that are democratically harmful, are also present in nuclear power.

Nuclear is also not a green technology. It is environmentally harmful and produces waste that will be a burden on future generations. Even if you believe these issues are theoretically solvable and nuclear power could be potentially deployed under your chosen form of government in the future, deploying it now takes us further away from that better future.

All material used as nuclear fuel has its origins in uranium mining. Uranium mining is a major extractive industry, with all that entails. While some of the materials used in battery components, for example, could also be criticised on that basis, the wastes produced by uranium mining are frequently radioactive as well as toxic. While the quantities of material used as fuel in a nuclear power station are relatively small, the amount of ore that needs to be mined in order to produce it is 2,500 times greater. This is due to the concentration of uranium in the ore, the amount of excess material discarded during enrichment, and the relative scarcity of the more reactive isotopes within uranium. Around 41% of uranium worldwide is extracted using the in-situ leaching method, a similar process to fracking.

Most troubling is the way that the harmful effects of uranium mining, like the testing of nuclear weapons, are so directly bound up with colonial geographies: again, nuclear power emerges from and then reproduces international relationships and systems. From draining spring water for mining concessions on contested Aboriginal land in Australia, to deaths from kidney failure and cancer in the Navajo nation in the US and radioactive water pollution in Nigeria, uranium mining has frequently visited its worst impacts upon colonised peoples who have seen none of the supposed benefits of nuclear power. I don't believe this callous disregard for certain human lives is a coincidental feature of the industry. Nuclear power is a direct outgrowth of the most extreme example of state violence: nuclear weapons. It should be no surprise that it also exhibits characteristics of colonialism as well. It is not correct to view nuclear power and nuclear weapons as separate, if linked, technologies. They are instead just two different applications of the same technology.

Nuclear power and nuclear weapons

The major industrial processes used for nuclear power generation are also used to produce nuclear weapons. This was quite literally the case in the UK for most of its twentieth century. The process of enriching uranium to make nuclear fuel only needs to be prolonged in order to enrich the uranium to weapons-grade. Nuclear reactors produce plutonium, and all of the early large reactors were created specifically for this purpose. Calder Hall, the reactor at Sellafield which is claimed as the first civil reactor in the world, was actually primarily used to produce plutonium for the UK weapons programme, although this was hidden from the public for years.

Whilst nuclear power had been theoretically proposed years before and was already being developed to power nuclear submarines, its use in civil electricity generation arose directly from Eisenhower's "Atoms for Peace" speech at the United Nations in 1951. However, the championing of nuclear power was not the main purpose of the speech and was in fact a late addition. The "Atoms for Peace" speech was the culmination of a year-long media campaign intended to convince the US public to accept the permanent presence of nuclear weapons and the Cold War standoff with the USSR. Original drafts of the speech were considered too bleak, so the proposals for global support for nuclear power were included to inject a note of hope.[13]

Nuclear power can only be realised through the violence and centralised control of the state forms we see in mid and late modernity. As we can see from nuclear power's current impasse in the UK, only the massive resources of the state are adequate to build the infrastructure and take on board the risks inherent to construction and waste management. The overlap between nuclear power and nuclear weapons can be seen very clearly in the example of the UK. The role of government funding for SMRs in subsidising Rolls-Royce's submarine reactor production is just one example among many that support for the UK's nuclear weapon programme is a major factor driving the government's enthusiasm for nuclear power. This supporting role, which is acknowledged at the highest level, is only possible because the two sectors draw from a common pool of knowledge and share a supply chain.

Preventing the spread of that technology and knowledge, and the destructive potential of materials and technologies, does not just necessarily imply secretive corporate entities, but also the existence of some kind of armed force to protect sites and materials. The UK actually has a separate police force for protecting the nuclear power sector, one of only three where officers are routinely armed. In 2021, does anyone really think it's a good idea to tackle climate change with a technology that would have required us to invent cops if they didn't already exist?

These problems are mirrored on the international scale. A world with widespread nuclear power is one that will live in permanent fear of nuclear weapons proliferation and where peaceful cooperation and coexistence will be harder to achieve. Every known case of proliferation has been disguised with a nuclear power programme that was claimed to be solely civil, and in many cases the technologies involved were taken directly from (actual or claimed) civil programmes elsewhere in the world.

The uranium enrichment centrifuge designs exported by A.Q. Khan to Iran, North Korea and Libya came from Urenco, the British/Dutch/German enrichment company. The design of the North Korean nuclear reactor at Yongbyon which supplied the plutonium for the country's nuclear weapons is based on the UK's Magnox reactor design that was used at Calder Hall and elsewhere.[14] The Syrian nuclear reactor bombed by Israel in 2007 was almost certainly derived from the same design. We do not need to endorse the US hegemonic taxonomy of good and bad states to recognise that a world where any government can build nuclear weapons will be unstable and dangerous. If we allow the UK to develop new nuclear power, we either have to deny other countries with fewer potential renewable power sources and greater development needs the right to also develop nuclear power, or be prepared for a world where the likes of Duterte and Mohammed bin Salman can achieve nuclear weapons capability at a whim.

Nuclear waste

Nuclear power has no solution to the problem of waste. The industry's preferred approach is to bury high level waste (the most radioactive and dangerous category) deep underground and forget about it. At the moment there is no working deep repository for high level waste anywhere in the world. Finland and Sweden have got the furthest towards building them, but their experience is not reassuring. Sweden recently discovered that the copper containers they were planning to use corroded much faster during experiments than they had anticipated. The official UK timeline for the Geological Disposal Facility (GDF) is planned around it being ready to accept waste in the 2040s, but nobody in the industry believes that is realistic.

The actual purpose of dates like this, and of the GDF plan in general, isn't that anyone thinks they will be fulfilled in anything like their current form – it's that the existence of these plans means that nobody today has to grapple with a problem that is actually insoluble. The plans are a convenient fiction disguising the fact that the problem is being constantly deferred for someone else to deal with. All that matters is to kick the can a little further down the road.

There are no specifications for the UK GDF yet, but when they emerge they will probably be along the following lines: no radioactive material should escape for around 500 to 1,000 years, and then only very small "safe" amounts would be able to escape until around 100,000 years have passed. These timescales make for interesting philosophical and ethical quandaries around how to construct warning signs and social or political forms that can transmit the necessity of leaving the waste alone for so long. However, the entire concept of deep geological "disposal" relies on the assumption that the people of the distant future will either have confidence in the technical solutions we choose to contain the waste or will be living in a civilisation so far advanced that putting in place additional measures will be inconsequential. In practice, neither of these assumptions should be taken for granted.

If our distant descendants are still around in 1,000 or 100,000 years and still retain any memory of our radioactive legacy, will they have any faith in infrastructure that is as distant to them as the reign of King Cnut or twice as old as the earliest cave paintings are to us?[15] The idea is even more absurd than the notion that having to deal with the waste will by then be a negligible burden. It's a laughable proposition, and we could learn a lot from the fact that the people who think otherwise are widely held to be pragmatists.

In truth, the radioactive wastes that we have currently generated, and whatever wastes are generated by the power plants being built now, will either endanger our descendants or place a burden of custodianship upon them that we have absolutely no right to impose. The financial costs of doing so are incalculable and it could never be ethically justifiable to pass them onto people who have no ability to consent.

Conclusion

In summary, nuclear power is antithetical to the world we want to see. From its origins as a figleaf to distract us from the grim truth of mutually assured destruction, to its recent resurrection as a bogus solution to climate change, it is inherently bound up with violent state forms and paranoid and secretive hierarchies. It cannot be deployed at a speed and scale to make a difference to climate change, but it will make the world less safe and stable at a time when we can least afford to manage the many problems that come with it. People will already have to deal with its legacy for countless generations and the only moral course of action is to decline to add to their burden by generating more waste.

Climate change mitigation measures need to be prefigurative of the other changes we want to see in the world. Technology will never be the solution to climate change, but any viable solution will need to deploy it alongside social change. Nuclear cannot deliver on even the limited grounds where it claims to make a difference and is a distracting dead end. In political circumstances where social change is not immediately realisable, we need to be advocating for technologies which are in harmony with the changes we want to see, not providing free PR for an industry which should have been left to die decades ago.

Democratically controlled renewable power generation is much more amenable to the types of adaptation and demand matching that make a zero carbon grid a realistic possibility. Renewables are less complex than nuclear power, much quicker and easier to deploy, and much more scalable. The technologies can easily be shared globally, and building more cross-border grid interconnectors will make it much easier to manage the variance of renewable generation. Rather than reproducing existing oppressive structures and relationships, these technologies are at the very least compatible with the relationships and institutions we would want to see as socialists.

Locally owned and run renewables, linked together in a web of global interdependencies, is exactly the kind of prefigurative solution that we need to be working towards, and it is actually cheaper and more realistic than nuclear power. Decarbonising electricity generation is the low hanging fruit of climate change adaptation, but if we carry it out in the right way, it will be easier to work towards just and equitable solutions in future steps. Nuclear has already blown its chance to be a meaningful part of that future – the only question is how quickly people on the left will recognise this, and how much more we are going to continue storing future problems by trying to resist its inevitable demise.

Notes

1. Even the government's inadequate plans call for 'an overwhelmingly decarbonised power system in the 2030s'. See "In-Depth Q&A: How Does the UK's "Energy White Paper" Aim to Tackle Climate Change?'" Carbon Brief. 16 December 2020.

2. As this piece was being prepared for publication it emerged that there has been a radiation leak in one of these reactors since October 2020, a little over a year since it began operation This further diminishes the likelihood that more EPR reactors will be built in the UK.

3. See: "When Investors Need to Restate Liabilities," The Footnotes Analyst. 31 March 2019. Undiscounted liabilities are listed as being £82bn. "Electricite de France Market Cap," YCharts puts market capitalisation at £43.5bn on 20 May 21.

4. Discounting is an accountancy practice where costs you will incur in the future are counted as literally costing less because you don't have to pay them yet. When used in combination with the massive timescales involved in nuclear decommissioning, discounting can produce interesting results.

5. There are other III+ generation reactor designs in Russia and India but these have not been seriously proposed for construction in Europe or North America and would probably need substantial adaptations to meet regulatory standards here.

6. See: Assessment of Reactors. Office for Nuclear Regulation. The Environment Agency consultation is due to publish in early 2022.

7. While Blair is widely believed to have personally driven the decision-making, he was far from alone in his enthusiasm. This surprisingly good BBC News article outlines the myriad links between the nuclear industry and New Labour: "Labour and the Nuclear Lobby". 23 May 2007.

8. In the event of an accident, operator liabilities are capped, regardless of their culpability. In 2017 the cap was raised from £140m to €1.2bn. See "Changes to the UK Nuclear Liability Regime: Implications for the Industry", Clifford Chance LLP.

9. The original plan was for operators to be charged a fixed fee by the government for waste disposal, transferring the risk to the state. The industry ran a successful lobbying campaign to ensure that fees would be low and operator risks minimised. See: "Waste and Decommissioning Financing Arrangements". n.d. No2NuclearPower.; Tim Webb. 2010. "EDF Ran Secret Lobbying Campaign to Reduce Size of Nuclear Waste Disposal Levy". The Guardian. 2 June 2010.

10. Base overnight cost per kW of Nuclear (light water reactor) is estimated at $6,034 vs. $1,248 for Solar PV with tracking in "Cost and Performance Characteristics of New Generating Technologies, Annual Energy Outlook 2021". 2021. US Energy Information Administration. Presumably without tracking technology the cost of installing solar would be even lower.

11. The most recent example in the UK is Sizewell B: Emily Gosden. "Sizewell B Nuclear Plant Forced to Stay Shut over Safety Concerns". The Times. 17 May 2021. Problems can cause plants to stay offline for years, or even close. In 2020 EDF announced it was to close Hunterston entirely after a two year closure. EDF hopes to be allowed to run the reactors for another few months, but the whole plant has essentially been retired due to a fault. See: "Scottish Nuclear Power Station to Shut down Early after Reactor Problems". *The Guardian.*. 27 August 2020.

12. Elaine Scarry. 2014. *Thermonuclear Monarchy: Choosing between Democracy and Doom.* New York: W.W. Norton & Company.

13. Stephanie Cooke. 2009. *In Mortal Hands: A Cautionary History of the Nuclear Age.* New York: Bloomsbury, pp.107-112.

14. Siegfried S. Hecker, Sean C. Lee, and Chaim Braun. 2010. "North Korea's Choice: Bombs Over Electricity". The Bridge 40 (2), pp.5–12.

15. Cnut's reign in England lasted from 1016 to 1035, just under a millenia ago. The oldest cave painting is dated to 45,000 years ago.

The Spokesman

Journal of the
Bertrand Russell Peace Foundation

Edited by Tony Simpson and Tom Unterrainer

Featuring independent journalism, opinion and analysis on peace and nuclear disarmament, human rights and civil liberties, and contemporary politics.

Subscribe ...
Subscription rates are (for three issues):
Individual subscriptions: £20
Individual subscriptions international: £25
Institutional subscriptions:
UK £33 Europe £38 RoW £40

Visit:

www.spokesmanbooks.org